A CHECKLIST OF THE PLANTS OF BUCKINGHAMSHIRE

(INCLUDING MILTON KEYNES & SLOUGH)

by
Roy Maycock & Aaron Woods

Milton Keynes Natural History Society

A CHECKLIST OF THE PLANTS OF
BUCKINGHAMSHIRE

INTRODUCTION

George Claridge Druce published his 'Flora of Buckinghamshire' in 1926. Since then, there has been no published work on the whole county. Various surveys have been carried out, notably that between about 1965 and 1985 when all tetrads were covered. Work has also been done in connection with other recording projects: the BSBI Monitoring Scheme, the BSBI Local Change Scheme, the 'Milton Keynes More Than Concrete Cows' book, the 'New Atlas of the British and Irish Flora', and the 'Vice-County Census Catalogue'. It was the publication of the latter and that of the 'Checklist of the Plants of Derbyshire' that prompted the compilation and publication of this book.

Buckinghamshire, as a relatively small, south midlands county of England, has ancient origins and its boundaries, as shown on a map of 1844, were used in its designation as a vice-county (number 24). The area covered here is that of the vice-county together with parts annexed from other counties e.g. Stokenchurch from Oxfordshire. There have been several minor changes to the administrative county boundaries relatively recently and two major ones: Slough etc. to Berkshire (though no longer an administrative county itself!) in 1974 and Milton Keynes becoming a unitary authority in 1997.

This publication is essentially a list of those vascular plants which have ever been recorded growing in the wild in Buckinghamshire, so it relies on the work of many recorders - both past (especially Druce) and present. However, to set the scene as it were, a brief account of some physical features is given.

Geology

The main axis of Buckinghamshire runs from north to south for about 85km whilst the maximum width is less than half this distance. In contrast, the solid geology outcrops run roughly northeast to southwest across the county, but may not be exposed. Essentially, there are clays, limestones (including chalk) and sands and gravels.

The oldest rocks are in the north, the youngest in the south. Small areas of Lias Clay in the extreme north are followed to the southeast by the Blisworth Limestone and rubbly limestone and clays that form the Cornbrash. Extensive areas of Oxford Clay follow, and south of these there is a complex of Kimmeridge Clay, capped in places with Portlandian and Purbeckian calcareous beds with occasional, higher areas of Lower Greensand. To the west is a small outcrop of Corallian beds, here the siliceous Arngrove Stone. Centrally in this part of the county are patches of Lower Cretaceous Sands and in the east a larger area of Lower Greensand, sometimes iron-rich, forming the Brickhill escarpment contrasting with the northern clay plains. The Gault Clay comes up to the base of the chalk of the Chiltern escarpment, but with a sliver of Upper Greensand in the southwest.

The Upper Chalk forms the bold escarpment of the Chiltern Hills that virtually separates the north of the county from the south. Together with the Middle Chalk and Lower Chalk a very large area is covered. The southeast 'triangle' of the county has isolated patches of Reading Beds of clays and sands, with London Clay completely overlain by drift deposits.

It is the rocks of the solid geology that form the framework on which the soils are built. Sometimes, where the rocks are near the surface, the soils are derived directly from them and influence the vegetation. In other cases, drift deposits are more important. Much of the north has Boulder Clay (covering the limestone and other clays) and some glacial gravels, all of which contain calcareous material. South of the Chiltern escarpment much of the dip slope is covered with Clay-with-Flints. This tends to give soils less alkaline in nature than those derived from the calcareous substrata. The large river valleys have alluvial deposits that are most extensive about the Ouse and Thames, the latter where Terrace Gravels extend several miles from the present river.

The effect of these formations is that most of the county has alkaline (or neutral) soils but the Brickhills (in the north-east) and the southeast triangle have acid soils. Consequently, it is generally the plants of acid soils that are our rarities.

Drainage

The two main river systems in the county are those of the Great Ouse (draining to the north) and the Thames (draining to the south). Each has an extensive tributary network, notably the Claydon Brook and Rivers Tove and Ouzel in the north and the Rivers Ray, Thame, Chess, Misbourne, Wye and Alderbourne to the south. The Chiltern scarp is dry!

Altitude and Climate

The county is part of lowland Britain. The highest point, near Wendover, is only at 260m, the lowest, where the Thames leaves the county, at about 15m. The climate is equable with no extremes of temperature or rainfall. Winds are generally light to moderate.

HABITATS

Much of Buckinghamshire is intensively farmed or under urban development and roads. However, a range of natural or semi-natural habitats remains to contribute to the well-over 2000 taxa recorded within this book.

Woodlands and Hedgerows

The county has large areas of ancient semi-natural woodlands (e.g. the beech woods of The Chilterns) or smaller remnants of larger, ancient forests (e.g. the Claydon Woods as part of the ancient Bernwood Forest) as well as plantations - sometimes of broad-leaved native trees, but often of conifers. Each type of woodland has its own characteristic species and some of the county, or even national rarities *e.g. Orchis militaris* (Military Orchid) are found in Chiltern woods.

Hedgerows form boundaries to woodlands, fields and highways. Some (assart hedges) are ancient remnants of felled woodland and so are richer in species than those more recently planted.

Water

The main rivers have been mentioned above but canals are also important features with their marginal and aquatic floras. The Grand Union Canal in Buckinghamshire is mainly in the north and east but has arms leading to Aylesbury and to Wendover (though this is no longer connected to the main canal). In the south the Slough arm joins the main canal just over the Hertfordshire border at Hillingdon.

The number of ponds is now drastically reduced but a few in fields still remain. The only really large bodies of water are man-made, like the water-filled gravel and clay pits, the balancing lakes in Milton Keynes or the canal top-up reservoirs. Consequently, the aquatic flora of the county is not as rich as in many others.

Grassland

Very little old grassland remains, but what does has some of our most interesting plants. The dry, chalk grasslands are rich in species and some have magnificent displays of orchids in their season. The Fringed Gentian (*Gentianella ciliata*) can still be seen at its only British location near Wendover where it was first seen in 1875. At the other extreme are a few wet meadows and fens, again rich in orchids but without real rarities. The only acid, boggy areas are in the south of the county (as at Black Park, Stoke Common and Burnham Beeches) and they have some of our rarest plants e.g. *Erica tetralix* (Cross-leaved Heath) and *Anagallis tenella* (Bog Pimpernel). Plants like *Parnassia palustris* (Grass-of-Parnassus) and *Narthecium ossifragum* (Bog Asphodel) have long since gone. The only acid soils in the north are around The Brickhills.

Urban Greenspace

Within the built-up areas is a wide range of micro-habitats suitable for plants to grow *e.g.* pavement cracks, walls, flower beds, sports fields and, probably most importantly, churchyards and other burial grounds. In the past, tips were a rich source of strange plants - many coming from the metropolis by barge along the Slough arm of the canal! It is often in any of these habitats that alien plants gain a foothold.

THE CHECKLIST

The list of plants is arranged alphabetically by scientific (Latin) names, mostly following Stace (1997). Taxa below subspecies level are not included. Where possible common (English) names appear in the second column. The information in the third column varies but gives some indication of national and/or local significance of the taxon, its status within the county and something of its distribution.

Both native and alien taxa are included in the list. Native taxa are those believed to be present in Buckinghamshire entirely due to natural processes. Alien taxa are those that have been introduced to Buckinghamshire by human activity. Those long established in Great Britain (*i.e.* before 1500) are archaeophytes, those established since 1500 are neophytes. Other aliens are listed as casual if they do not maintain themselves in the county by seed or vegetative means, or are obviously planted.

Information about the taxa within the county can also be gleaned from the font type and size:

Key (Column 1)

BOLD FACE CAPITALS	**NATIVE (POST 1986 RECORD)**
Bold face lower case	**Native (1970 - 1986 record)**
LIGHT FACE CAPITALS	ARCHAEOPHYTE
Light face lower case	Neophyte
Light face italics	*Casual*
*	Pre-1970 record

Key (Column 3)

During the compilation of this Checklist, the Joint Nature Conservation Committee (JNCC) published a new Red Data List for Vascular Plants in Great Britain. Previous Red Data Books and the Scarce Plants book accorded rarity status to plants depending on their hectad (10km square) frequency. The new list uses criteria set out by The International Union for the Conservation of Nature and National Resources (IUCN) and relates to the conservation status of the plants. It includes threatened taxa as well as those (and there are many in Bucks!) that are of 'Least Concern' and will be used to inform future priority setting in the Biodiversity Action Plan (BAP) process. We have not considered this and have retained the spatial frequency for determining rarity within the county. Categories included are:

EX	Extinct
CR	Critically Endangered
EN	Endangered
VU	Vulnerable
NT	Near Threatened

together with:

NS	Nationally Scarce
RDB	British Red Data Book

and:

Dates	Latest recorded date where considered significant
Pre-1926	Record in Druce's Flora, but no precise date given

ACKNOWLEDGEMENTS

As two members of Milton Keynes Natural History Society, we are pleased that the Society has taken the responsibility for publishing this Checklist. Without the help of other members of the Society and people from around the county and beyond the list would not have been so complete. Special thanks must go to the BSBI Referees who have vetted and advised on *Hieracium* (David McCosh), *Rosa* (Tony Primavesi and Roger Maskell), *Rubus* (Alan Newton) and *Taraxacum* (John Richards). Alan Showler, Chris Boon or Robert Williams have read the list and text at various stages in its compilation and made valuable suggestions. Alan Showler has also made a huge contribution to the records, especially from the High Wycombe area, adding to those already made by the late Betty Marcan - a stalwart recorder. Charlotte Matthews was responsible for the artwork. The staff at the Buckinghamshire and Milton Keynes Records Centre have been most generous and provided us with invaluable support. Nick Moyes at the Derbyshire Museum was most helpful and encouraging during the early stages of the work.

We have also been fortunate to receive financial support from:

Buckinghamshire County Council
Aylesbury Vale District Council
South Bucks District Council
Milton Keynes Council
The Chilterns Conservation Board
Thames Valley Enviromental Records Centre
Berkshire, Buckinghamshire & Oxfordshire Wildlife Trust (Chilterns Region)
Milton Keynes Natural History Society (Gordon Osborn Bequest Fund)
Alan Showler
The estate of the late Robert Raper

POSTSCRIPT

This list has been produced to the best of our abilities and has relied not only on our own records but those made by many other people. To those still living we offer our thanks; to those no longer with us we must be grateful that they recorded in Buckinghamshire.

What can you do?

If, after having consulted the list, you discover errors, please let us know.

If you ever visit Buckinghamshire on a botanising trip or just happen to make casual records, please let us know. Very few records ever get to us from outside the county and, even within it, there are only a few people who let us know what they see. For example: in the 1970s it was surprising to find how few plant records there were in the tetrad in which *Epipogium aphyllum* (Ghost Orchid) grew! How many people had been to look for it and bothered about nothing else?

For each record we need to know:
the name of the taxon
the name of the recorder/determiner
the date on which the record was made
the name and grid reference of the site from which the record
was made
the habitat.

If you consider a voucher specimen to be necessary, please send one so that it can be deposited in the County Museum herbarium. But, note legal requirements for taking specimens!

Abies alba	European Silver-fir	Rare; in a few plantations
Abies grandis	Giant Fir	Rare; in a few plantations; *e.g.* Kilwick Wood
Abies nordmanniana	Caucasian Fir	Rare; planted in woods; *e.g.* Rushmere Park
Abies pinsapo	Spanish Fir	Very rare; planted in parkland; *e.g.* Aston Clinton
Abies procera	Noble Fir	Occasional; planted in woods & plantations
Abutilon theophrasti	Velvetleaf	1976 (Langley); disturbed ground
ACER CAMPESTRE	Field Maple	Very common; woods & hedgerows
Acer cappadocicum	Cappadocian Maple	Very rare; planted in parkland, rarely self-sown; *e.g.* Halton
Acer negundo	Ashleaf Maple	Rare; planted in parkland, churchyards & hedges
Acer platanoides	Norway Maple	Frequent; woods, parkland & hedges, often self-sown
Acer pseudoplatanus	Sycamore	Very common; woods & hedgerows
Acer saccharinum	Silver Maple	Rare; planted in parkland
Aceras anthropophorum *	Man Orchid	EN; *c.*1867 (near Wendover); calcareous grassland
Achillea filipendulina	Fern-leaved Yarrow	Rare; waste places
ACHILLEA MILLEFOLIUM	Yarrow	Very common; grassland & roadsides
ACHILLEA PTARMICA	Sneezewort	Scarce; riversides, fens & wet grassland
Aconitum napellus agg.	Monk's-hood	Very rare; woodland; *e.g.* Biddlesden Park
Acorus calamus	Sweet-flag	Rare; canal & riversides; decreasing
Acorus gramineus	Slender Sweet-flag	1998 (Singleborough); pond
Adiantum capillus-veneris	Maidenhair-fern	NS; very rare; garden centre weed, greenhouses & walls
ADONIS ANNUA *	Pheasant's-eye	EN; pre-1926 (near Princes Risborough); arable fields
ADOXA MOSCHATELLINA	Moschatel	Scarce; damp woods & hedgerows
AEGOPODIUM PODAGRARIA	Ground-elder	Common; gardens & waste places; occasional in woods
Aesculus carnea	Red Horse-chestnut	Rare; planted in parkland, gardens & churchyards; rarely self-sown; *e.g.* Bradwell Common
Aesculus hippocastanum	Horse-chestnut	Common; parkland, roadsides & woods; sometimes self-sown
AETHUSA CYNAPIUM SSP. AGRESTIS *		Pre-1926; arable fields; probably overlooked
AETHUSA CYNAPIUM SSP. CYNAPIUM	Fool's Parsley	Very common; weed of disturbed ground
Ageratum houstonianum	Flossflower	Very rare; of waste places; *e.g.* Clifton Reynes & Wycombe
AGRIMONIA EUPATORIA	Agrimony	Common; rough grassland & wood margins
AGRIMONIA PROCERA	Fragrant Agrimony	Very rare; grassland; possibly overlooked
AGROSTEMMA GITHAGO	Corncockle	Extinct; arable fields; now often planted in wild flower mixes
AGROSTIS CANINA	Velvet Bent	Rare; damp acidic grassland
AGROSTIS CAPILLARIS	Common Bent	Locally common; dry grassland, often on acid soils
Agrostis castellana	Highland Bent	1981 (New Wavendon Heath); from grass seed mix
Agrostis exarata	Spike Bent	1981 (New Wavendon Heath); from grass seed mix
AGROSTIS GIGANTEA	Black Bent	Uncommon; arable fields & waste places on light soils
AGROSTIS STOLONIFERA	Creeping Bent	Very common; grassland & wet places
AGROSTIS VINEALIS	Brown Bent	Rare; dry sandy places; probably under-recorded
Ailanthus altissima	Tree-of-heaven	Rare; planted in churchyards & urban areas; occasionally self-sown
AIRA CARYOPHYLLEA	Silver Hair-grass	Very rare; dry heathy places & railway ballast
AIRA PRAECOX	Early Hair-grass	Rare; dry open places
AJUGA REPTANS	Bugle	Common; woods & damp grassland
Alcea rosea	Hollyhock	Rare; waste places, walls & pathsides
ALCHEMILLA FILICAULIS SSP. VESTITA	Hairy Lady's-mantle	Rare; woodland rides & damp grassland
Alchemilla mollis	Garden Lady's-mantle	Increasing; waste places near habitation; frequently self-sown
ALCHEMILLA XANTHOCHLORA	Intermediate Lady's-mantle	Rare; woodland rides & grassland
ALISMA LANCEOLATA	Narrow-leaved Water-plantain	Rare; canal & pondsides
ALISMA PLANTAGO-AQUATICA	Water-plantain	Common; river, canal & pondsides
ALLIARIA PETIOLATA	Garlic Mustard	Very common; woodland margins & hedgerows
Allium cepa	Onion	1991 (Aylesbury); relic of cultivation on allotments
Allium moly	Yellow Garlic	1980 (Burnham Beeches); rubbish heap
Allium oleraceum *	Field Garlic	VU; 1954 (Bourne End); dry grassy places & hedges
Allium paradoxum	Few-flowered Garlic	Rare but increasing; roadsides & churchyards
Allium sativum	Garlic	Very rare; waste places
Allium schoenoprasum	Chives	Very rare; waste places
Allium subhirsutum	Hairy Garlic	Very rare; parkland & roadsides; *e.g.* Halton
Allium triquetrum	Three-cornered Garlic	Very rare; parkland & roadsides; *e.g.* Black Park
ALLIUM URSINUM	Ramsons	Rare; damp woods, mostly on calcareous soils
ALLIUM VINEALE	Wild Onion	Uncommon; roadsides & churchyards; especially in the south
Alnus cordata	Italian Alder	Rare; planted in parkland & by roadsides
ALNUS GLUTINOSA	Alder	Uncommon; by rivers, streams & in damp woods
Alnus incana	Grey Alder	Scarce; planted in parkland & by roadsides
ALOPECURUS AEQUALIS	Orange Foxtail	Very rare; pond margins; mainly in the south

ALOPECURUS GENICULATUS	Marsh Foxtail	Common; wet grassland & pond margins
ALOPECURUS MYOSUROIDES	Black-grass	Common; arable fields & waste places
ALOPECURUS PRATENSIS	Meadow Foxtail	Very common; grassland
Alopecurus x haussknechtianus		1986 (Tiddenfoot Water Park); water-filled sandpit
Althaea hirsuta *	Rough Marsh-mallow	1968 (Princes Risborough); tip & pathside
Althaea officinalis	Marsh-mallow	Very rare; planted in grassy places; *e.g.* Great Linford
Alyssum alyssoides *	Small Alison	1865 (Little Marlow); cultivated fields
Alyssum saxatile	Golden Alison	Rare; persistent on walls; *e.g.* Loughton
Amaranthus blitum *	Livid Amaranth	*c.*1940 (Burcott); rubbish tips
Amaranthus caudatus *	Love-lies-bleeding	Pre-1926 (Langley & Iver); waste places
Amaranthus hybridus	Green Amaranth	2003 (Great Kingshill); disturbed ground
Amaranthus hypochondriacus	Prince's-feather	1998 (Bletchley); garden weed
Amaranthus quitensis	Mucronate Amaranth	1970 (Iver); rubbish heap
Amaranthus retroflexus	Common Amaranth	Rare; waste places
Amaranthus thunbergii *	Thunberg's Pigweed	1919 (Langley); waste places
Ambrosia artemisiifolia	Ragweed	2004 (Lane End); maize field
Ambrosia peruviana *	Peruvian Ragweed	1876 (Slough & Eton); cultivated land
Amelanchier lamarckii	Juneberry	Rare; woods & commons; *e.g.* Linford Wood
Ammi majus	Bullwort	1998 (High Wycombe); birdseed alien
Ammi visnaga	Toothpick Plant	1985 (Amersham); garden weed
Amsinckia lycopsoides *	Scarce Fiddleneck	1917 (Naphill); waste ground
Amsinckia micrantha	Common Fiddleneck	2002 (Wycombe Marsh); waste places
ANACAMPTIS PYRAMIDALIS	Pyramidal Orchid	Scarce; chalk grassland; recently spreading northwards
ANAGALLIS ARVENSIS SSP. ARVENSIS	Scarlet Pimpernel	Common; disturbed ground & grassy places
ANAGALLIS ARVENSIS SSP. FOEMINA	Blue Pimpernel	Very rare; disturbed ground & grassy places
Anagallis minima	Chaffweed	NT; 1972 (Stoke Common); sandy commons in the south
ANAGALLIS TENELLA	Bog Pimpernel	Very rare; bogs & fens; rediscovered 2002
Anaphalis margaritacea *	Pearly Everlasting	Pre-1926 (St Leonards & Newton Blossomville); hedges
ANCHUSA ARVENSIS	Bugloss	Very rare; disturbed sandy or gravelly ground; decreasing
Anchusa azurea *	Garden Anchusa	Pre-1926 (Iver); waste places
Anchusa officinalis *	Alkanet	*c.*1955 (Wooburn); rubbish tips & waste places
Anchusa undulata *	Undulate Alkanet	Pre-1926 (Slough); rubbish tip
Anchusa variegata	Variegated Alkanet	1982 (Haddenham); with soil imported from Rhodes
Anemone apennina	Blue Anemone	Very rare; woods & parkland
Anemone blanda	Greek Anemone	Very rare; roadsides & parkland; *e.g.* Bradwell
ANEMONE NEMOROSA	Wood Anemone	Uncommon; woods, especially on humus-rich soils
Anemone ranunculoides	Yellow Anemone	Very rare; park & churchyard; *e.g.* Drayton Beauchamp
Anemone x hybrida	Japanese Anemone	Very rare; waste places near habitation
Anethum graveolens	Dill	1972 (Gerrards Cross); rubbish tip
ANGELICA ARCHANGELICA	Garden Angelica	2002 (Hughenden Valley); waste places
ANGELICA SYLVESTRIS	Wild Angelica	Common; damp woods, ditches & watersides
Anisantha diandra	Great Brome	Rare; arable fields & roadside; *e.g.* Honeyburge; increasing
Anisantha rigida	Ripgut Brome	1997 (Rowsham); roadside
ANISANTHA STERILIS	Barren Brome	Very common; waste & cultivated ground
Anisantha tectorum	Drooping Brome	1970 (Stoke Common); waste ground
ANTHEMIS ARVENSIS	Corn Chamomile	EN; very rare; arable fields; decreasing
ANTHEMIS COTULA	Stinking Chamomile	VU; scarce; arable & waste ground
Anthemis ruthenica *	Russian Chamomile	Pre-1926 (Slough); waste ground
Anthemis tinctoria *	Yellow Chamomile	1969 (Boveney); railway banks & waste places
Anthoxanthum aristatum ssp. puelii	Annual Vernal-grass	1982 (New Wavendon Heath); from grass seed mix
ANTHOXANTHUM ODORATUM	Sweet Vernal-grass	Common; grassland on less alkaline soils
ANTHRISCUS CAUCALIS	Bur Chervil	2001 (Holmer Green); arable fields & walls
Anthriscus cerefolium	Garden Chervil	Very rare; waste places & roadsides
ANTHRISCUS SYLVESTRIS	Cow Parsley	Very common; woods, hedges, roadsides & churchyards
Anthyllis vulneraria ssp. polyphylla		Very rare; roadsides; *e.g.* Little Brickhill
ANTHYLLIS VULNERARIA SSP. VULNERARIA	Kidney Vetch	Uncommon; calcareous grassland & bare Oxford clay
Antirrhinum majus	Snapdragon	Uncommon; walls & waste places
APERA SPICA-VENTI	Loose Silky-bent	NT; NS; rare; sandy arable fields in the south
APHANES ARVENSIS	Parsley-piert	Uncommon; arable fields, short turf and urban areas
APHANES AUSTRALIS	Slender Parsley-piert	Rare; bare, sandy ground; may be under-recorded
APIUM GRAVEOLENS	Wild Celery	Very rare; marshy grassland; also a garden escape
APIUM INUNDATUM	Lesser Marshwort	Very rare; ponds on commons in south; decreasing
APIUM NODIFLORUM	Fool's-water-cress	Common; ditches & watersides
AQUILEGIA VULGARIS	Columbine	Very rare; woods; also a frequent garden escape

ARABIDOPSIS THALIANA	Thale Cress	Frequent; walls, banks & bare ground on light soils
Arabis caucasica	Garden Arabis	Uncommon; walls near habitation
Arabis glabra	Tower Mustard	EN; RDB; very rare; hedgebanks & sandy places
ARABIS HIRSUTA	Hairy Rock-cress	Very rare; bare, calcareous ground in The Chilterns
Aralia chinensis	Chinese Angelica-tree	Very rare; planted in gardens; rarely self-sown nearby; e.g. Conniburrow
Araucaria araucana	Monkey-puzzle	Rare; planted in churchyards & gardens
Arbutus unedo	Strawberry-tree	c.1980 (Newlands Park); planted
ARCTIUM LAPPA	Greater Burdock	Uncommon; waste places, especially near water
ARCTIUM MINUS	Lesser Burdock	Common; woods, hedgerows & waysides
ARCTIUM NEMOROSUM	Wood Burdock	Rarely recorded; doubtless overlooked
ARENARIA SERPYLLIFOLIA SSP. LEPTOCLADOS	Slender Sandwort	Rare; arable fields & walls
ARENARIA SERPYLLIFOLIA SSP. SERPYLLIFOLIA	Thyme-leaved Sandwort	Common; arable fields & walls
Aristolochia clematitis	Birthwort	Very rare; hedges & under trees; e.g. Soulbury allotments
Armeria maritima ssp. maritima	Thrift	Very rare; roadside at Prestwood
ARMORACIA RUSTICANA	Horse-radish	Frequent; roadsides & riverbanks; especially in the north
Arnoseris minima	Lamb's Succory	EX; 1972 (Ibstone); sandy fields
ARRHENATHERUM ELATIUS	False Oat-grass	Very common; grasslands & arable fields
ARTEMISIA ABSINTHIUM	Wormwood	Rare; waste places; decreasing
Artemisia annua *	Annual Mugwort	1927 (Iver); waste places
Artemisia biennis	Slender Mugwort	1977 (Langley); waste places
Artemisia verlotiorum	Chinese Mugwort	Very rare; rubbish tips & urban areas
ARTEMISIA VULGARIS	Mugwort	Common; roadsides & waste places
Arum italicum ssp. italicum	Italian Lords-and-Ladies	Very rare; parks, churchyard & riverbank
Arum italicum ssp. neglectum	Scarce Lords-and-Ladies	NT; very rare; riverbank near Huntsmoor
Arum italicum ssp. italicum x italicum ssp. neglectum		Very rare; riverbank near Huntsmoor
ARUM MACULATUM	Lords-and-Ladies	Very common; woods & hedges
Asarum europaeum	Asarabacca	Very rare; woods; e.g. Weston Turville
ASPARAGUS OFFICINALIS SSP. OFFICINALIS	Garden Asparagus	Scarce; waste ground
Asperugo procumbens *	Madwort	Pre-1926 (Iver); waste places
Asperula arvensis	Blue Woodruff	1991 (Aylesbury); disturbed ground
ASPERULA CYNANCHICA SSP. CYNANCHICA	Squinancywort	Rare; short turf on chalk soils
ASPLENIUM ADIANTUM-NIGRUM	Black Spleenwort	Scarce; walls; increasing
Asplenium fontanum *	Smooth Rock-spleenwort	Pre-1809 (Amersham); wall of church
ASPLENIUM RUTA-MURARIA	Wall-rue	Uncommon; wall interstices; increasing
ASPLENIUM TRICHOMANES SSP. QUADRIVALENS	Maidenhair-spleenwort	Uncommon; walls
Aster lanceolatus	Narrow-leaved Michaelmas-daisy	Rare; waste places
Aster novae-angliae	Hairy Michaelmas-daisy	1998 (High Wycombe); waste places
Aster novi-belgii	Confused Michaelmas-daisy	Very rare; waste ground; much over-recorded in the past
Aster tripolium *	Sea Aster	Pre-1926 (Taplow); rubbish tip
Aster x salignus	Common Michaelmas-daisy	Common; waste places & roadsides
Aster x versicolor	Late Michaelmas-daisy	Very rare; waste places & roadsides
ASTRAGALUS GLYCYPHYLLOS	Wild Liquorice	Rare; grassy places & under hedges
Astrantia major	Astrantia	1974 (near Fulmer); shady places
ATHYRIUM FILIX-FEMINA	Lady-fern	Uncommon; damp woods on peaty soils
Atriplex hortensis	Garden Orache	Very rare; disturbed ground; e.g. Soulbury allotments
Atriplex littoralis	Grass-leaved Orache	1999 (High Wycombe); waste places
ATRIPLEX PATULA	Common Orache	Common; cultivated ground & waste places
ATRIPLEX PROSTRATA	Spear-leaved Orache	Common; cultivated ground & waste places
Atriplex tatarica *	Tatarian Orache	1906 (Langley); rubbish tips
ATROPA BELLADONNA	Deadly Nightshade	Rare; open woods & scrub on calcareous soils
Aubrieta deltoidea	Aubretia	Frequent; walls near habitation
Aucuba japonica	Spotted-laurel	Common; planted in churchyards, shrubberies & gardens
Avena barbata *	Slender Oat	Pre-1926 (Slough Mill); waste ground
AVENA FATUA	Wild-oat	Common; arable fields

Avena sativa ssp. *sativa*	Cultivated Oat	Scattered; relic of cultivation
Avena sterilis ssp. *ludoviciana*	Winter Wild-oat	Rare; arable fields, especially on heavy soils
Avena sterilis ssp. *sterilis* *	Animated Oat	1959 (Wooburn); waste places
Avena strigosa *	Bristle Oat	Pre-1926 (Fenny Stratford); cornfield
Axyris amaranthoides *	Russian Pigweed	1917 (Naphill); waste ground
Azolla filiculoides	Water Fern	Rare; canals & rivers; populations fluctuate
Baldellia ranunculoides *	Lesser Water-plantain	NT; 1869 (Hyde Heath); bogs & ponds
BALLOTA NIGRA SSP. MERIDIONALIS	Black Horehound	Frequent; hedgerows & waste places
Barbarea intermedia	Medium-flowered Winter-cress	Rare; cultivated fields & waste places
Barbarea verna *	American Winter-cress	1918 (Denham); waste & cultivated land
BARBAREA VULGARIS	Winter-cress	Common; damp grassland by water or roadsides
Bassia scoparia	Summer-cypress	1972 (Gerrards Cross); rubbish tip
Beckmannia syzigachne *	American Slough-grass	1911 (near Uxbridge); waste places
Begonia x *semperflorens*	Wax Begonia	Very rare; waste places near habitation
Bellevalia romana *	Roman Squill	1931 (Eton); grassland
BELLIS PERENNIS	Daisy	Very common; short grassland *e.g.* lawns
Berberis darwinii	Darwin's Barberry	Scarce; planted in shrubberies; rarely self-sown
Berberis gagnepainii	Gagnepain's Barberry	Very rare; hedge at Lower Cadsden
Berberis julianae	Chinese Barberry	Rare; planted in shrubberies, rarely self-sown
Berberis thunbergii	Thunberg's Barberry	Rare; planted in gardens; rarely self-sown
BERBERIS VULGARIS	Barberry	Very rare; hedgerows; formerly more common
Berberis x *stenophylla*	Hedge Barberry	Rare; planted in shrubberies; rarely self-sown
Bergenia crassifolia	Elephant-ears	Very rare; commonland at Moorend Common
Berteroa incana	Hoary Alison	1982 (Denham); waste places & roadsides
BERULA ERECTA	Lesser Water-parsnip	Scarce; ditches, streams, less often by rivers or canals
Beta vulgaris ssp. *cicla*	Foliage Beet	1999 (Wexham); waste ground & allotments
Beta vulgaris ssp. *vulgaris*	Root Beet	Rare; old allotments & waste places
Betula jacquemontii	Jacquemont's Birch	Very rare; planted in parks & gardens; very rarely self-sown
BETULA PENDULA	Silver Birch	Common; woodland & heaths, especially on light soils
BETULA PUBESCENS SSP. PUBESCENS	Downy Birch	Uncommon; woodland, confined to heathy acid soils
BETULA X AURATA	Hybrid Birch	Very rare; may be present with parents; under-recorded
BIDENS CERNUA	Nodding Bur-marigold	Very rare; pond & lake margins, especially in the south
BIDENS CONNATA	London Bur-marigold	Scarce; banks of the Grand Union Canal
Bidens ferulifolia	Fern-leaved Beggarticks	Very rare; waste places near habitation
BIDENS FRONDOSA	Beggarticks	Rare, but increasing; lake and canal banks in the north
BIDENS TRIPARTITA	Trifid Bur-marigold	Scarce; by still or slow-moving water, decreasing
BLACKSTONIA PERFOLIATA	Yellow-wort	Scarce; on dry chalk or calcareous clay banks
BLECHNUM SPICANT	Hard-fern	Scarce; sparingly in woods & heaths on acid soils
Blysmus compressus	Flat-sedge	VU; 1972 (Dorney Common); wet meadows
BOLBOSCHOENUS MARITIMUS	Sea Club-rush	Very rare; gravel pits, probably originally introduced by waterfowl
Borago officinalis	Borage	Rare; waste places
BOTRYCHIUM LUNARIA	Moonwort	Very rare; believed extinct, but refound in N. Bucks SSSI in 1995
Brachyglottis 'Sunshine'	Shrub Ragwort	Rare; planted in parks & gardens
BRACHYPODIUM PINNATUM	Tor-grass	Rare; calcareous grassland & roadsides, very rarely in open woodland
BRACHYPODIUM SYLVATICUM	False Brome	Very common; woods & hedges, can also be invasive on chalk grassland
Brassica elongata ssp. *integrifolia* *	Long-stalked Rape	1898 (Water Eaton); waste places
Brassica juncea	Chinese Mustard	1995 (Gerrards Cross); possibly of birdseed origin
Brassica napus ssp. *oleifera*	Oil-seed Rape	Frequent; roadsides & waste places
Brassica napus ssp. *rapifera*	Swede	Pre-1985 (Langley Park); rubbish tip
BRASSICA NIGRA	Black Mustard	Scarce; roadsides, river banks, decreasing in the north
Brassica oleracea	Cabbage	Rare; waste places
BRASSICA RAPA SSP. CAMPESTRIS	Wild Turnip	Rare; river banks
Brassica rapa ssp. *oleifera*	Turnip-rape	*c.*1995 (Heelands); garden weed
Brassica rapa ssp. *rapa*	Turnip	Rare; crop relic
Briza maxima	Greater Quaking-grass	Very rare; waste places
BRIZA MEDIA	Quaking-grass	Common; old grassland especially on calcareous or neutral soils

Briza minor	Lesser Quaking-grass	1976 (Fingest); waste places
BROMOPSIS BENEKENII	Lesser Hairy-brome	NS; rare; woods in the Wycombe area
BROMOPSIS ERECTA	Upright Brome	Frequent; dry calcareous grassland
Bromopsis inermis ssp. inermis	Hungarian Brome	Very rare; roadsides & arable fields
BROMOPSIS RAMOSA	Hairy-brome	Common; woods & shady hedges on damp soils
Bromus arvensis	Field Brome	1979 (Brill Common); arable fields & waste places
Bromus brachystachys *	Palestine Brome	Pre-1926 (Slough); waste places
Bromus briziformis *	Rattlesnake Brome	1903 (near Langley); rubbish tips
BROMUS COMMUTATUS	Meadow Brome	Common; rough grassland or meadows
BROMUS HORDEACEUS SSP. HORDEACEUS	Soft-brome	Very common; rough ground & arable fields
BROMUS HORDEACEUS SSP. LONGIPEDICELLATUS	Rare; grassy places; *e.g.* Singleborough; probably overlooked	
Bromus interruptus *	Interrupted Brome	EX; 1913 (Princes Risborough); cultivated fields
Bromus japonicus *	Thunberg's Brome	Pre-1926 (Slough & Iver); waste places
Bromus lepidus	Slender Soft-brome	1987 (Pulpit Hill); grassland
BROMUS RACEMOSUS	Smooth Brome	Rare; grassland; formerly more widespread
Bromus scoparius *	Broom Brome	Pre-1926 (Slough & Taplow); waste places
BROMUS SECALINUS	Rye Brome	VU; very rare; arable fields; formerly more common
Bromus squarrosus *	Rough Brome	Pre-1926 (Hanslope & Slough); waste places
Bromus x pseudothominei	Lesser Soft-brome	Rare; rough ground; probably under-recorded
Brunnera macrophylla	Great Forget-me-not	Very rare; hedges; may persist for a few years
BRYONIA DIOICA	White Bryony	Common; hedgerows & woodland edges
Buddleja alternifolia	Alternate-leaved Butterfly-bush	1984 (Latimer); churchyard
Buddleja davidii	Butterfly-bush	Common; waste ground, walls & railways in urban areas; increasing
Buddleja globosa	Orange-ball-tree	Very rare; waste places near habitation
Bunias erucago *	Southern Warty-cabbage	Pre-1926 (Slough); waste places
Bunias orientalis	Warty-cabbage	Rare; roadsides & hedgebanks in the south
BUNIUM BULBOCASTANUM	Great Pignut	RDB; very rare; chalk hills near Ivinghoe
Bupleurum rotundifolium	Thorow-wax	1981 (Burnham); cornfields
Bupleurum subovatum	False Thorow-wax	1985 (Amersham); waste ground
BUTOMUS UMBELLATUS	Flowering-rush	Scarce; canal & river sides, less often in ponds (where it may be planted)
BUXUS SEMPERVIRENS	Box	RDB; very rare; chalk coombes; also frequently planted
CALAMAGROSTIS CANESCENS	Purple Small-reed	1989 (Leckhampstead Wood); woodland ride
CALAMAGROSTIS EPIGEJOS	Wood Small-reed	Frequent; woods, open clay areas & waste places; apparently increasing
Calendula arvensis *	Field Marigold	1918 (Denham); waste ground
Calendula officinalis	Pot Marigold	Rare; waste places
Calla palustris	Bog Arum	1999 (Commonhill Wood); ponds
Callistephus chinensis *	China Aster	Pre-1926 (Iver); waste places
Callitriche brutia *	Pedunculate Water-starwort	Pre-1926 (Dropmore); ponds
CALLITRICHE HAMULATA	Intermediate Water-starwort	2003 (Whitchurch); streams, ponds & gravel pits
CALLITRICHE OBTUSANGULA	Blunt-fruited Water-starwort	Very rare; rivers & ponds; *e.g.* Olney Park
CALLITRICHE PLATYCARPA	Various-leaved Water-starwort	Scarce; rivers & ponds
CALLITRICHE STAGNALIS	Common Water-starwort	Frequent; ponds, streams & wet woodland rides
CALLUNA VULGARIS	Heather	Scarce; dry heathy soils; suffering through lack of management
Calocedrus decurrens	Californian Incense-cedar	1970 (Green Park); parkland; planted
CALTHA PALUSTRIS	Marsh-marigold	Uncommon; ponds, streams & wet meadows; decreasing
Calystegia pulchra	Hairy Bindweed	Very rare; roadsides & hedgerows; *e.g.* Naphill
CALYSTEGIA SEPIUM SSP. SEPIUM	Hedge Bindweed	Very common; wood margins, hedges & waste places
Calystegia silvatica	Large Bindweed	Scarce; hedges & waste places; often near habitation
Calystegia x lucana	Hybrid Bindweed	1995 (Iver); hedge
Camelina alyssum *	Stinking Flaxweed	1869 (Castlethorpe); railwayside
Camelina sativa	Gold-of-pleasure	1970 (Chesham); cultivated fields
CAMPANULA GLOMERATA	Clustered Bellflower	Scarce; open calcareous grassland
CAMPANULA LATIFOLIA	Giant Bellflower	Rare; woods in the north

Campanula latiloba	Delphinium Bellflower	Pre-1985 (South Heath); parkland & waste places
Campanula medium	Canterbury-bells	2000 (Downley); waste places
Campanula persicifolia	Peach-leaved Bellflower	Uncommon; walls & waste places
Campanula portenschlagiana	Adria Bellflower	Very rare; walls
Campanula poscharskyana	Trailing Bellflower	Frequent; walls & pavement cracks
Campanula rapunculoides	Creeping Bellflower	Rare; hedges & waste places
CAMPANULA RAPUNCULUS	Rampion Bellflower	EN; 2004 (Chearsley & Bletchley); roadsides & lawns
CAMPANULA ROTUNDIFOLIA	Harebell	Uncommon; dry grassland & heaths
Campanula takesimana	Korean Bellflower	1998 (North Crawley); between paving
CAMPANULA TRACHELIUM	Nettle-leaved Bellflower	Locally common; woods from The Chilterns southward
Cannabis sativa	Hemp	Rare; waste & cultivated ground; often from birdseed or fishing bait; sometimes planted
CAPSELLA BURSA-PASTORIS	Shepherd's-purse	Very common; gardens, waste ground & arable fields
CARDAMINE AMARA	Large Bitter-cress	Very rare; shady streams & canals, mainly in the south
CARDAMINE BULBIFERA	Coralroot	NS; Locally common; Chiltern woods speciality; planted elsewhere
CARDAMINE FLEXUOSA	Wavy Bitter-cress	Common; damp shady places; increasing
CARDAMINE HIRSUTA	Hairy Bitter-cress	Common; gardens, walls & waste places; often spread in plant pots
CARDAMINE PRATENSIS	Cuckooflower	Very common; damp grassy places & churchyards
CARDUUS CRISPUS SSP. MULTIFLORUS	Welted Thistle	Very common; roadsides & sheltered hedges
CARDUUS NUTANS	Musk Thistle	Uncommon; open calcareous grassland
Carduus tenuiflorus	Slender Thistle	1977 (Calverton); ballast alien
CARDUUS X STANGII		1999 (Hughenden); grassy places
CAREX ACUTA	Slender Tufted-sedge	Rare; mostly in the larger rivers
CAREX ACUTIFORMIS	Lesser Pond-sedge	Uncommon; rivers, canals, ponds & wet woodland
Carex appropinquata *	Fibrous Tussock-sedge	NT; 1925 (Denham); marsh
CAREX BINERVIS	Green-ribbed Sedge	Very rare; heathy places; especially in the south
CAREX CARYOPHYLLEA	Spring Sedge	Rare; short turf, especially on chalk; decreasing
CAREX CURTA	White Sedge	Very rare; boggy woodland; only at Black Park
Carex diandra	Lesser Tussock-sedge	NT; 1973 (Blackend Spinney); wet woodland
CAREX DISTANS	Distant Sedge	Very rare; wet rushy grassland; previously overlooked
CAREX DISTICHA	Brown Sedge	Rare; wet meadows & marshes
CAREX DIVULSA SSP. DIVULSA	Grey Sedge	Uncommon; roadsides, hedges & churchyards; often on calcareous soils
CAREX DIVULSA SSP. LEERSII	Many-leaved Sedge	Rare; hedges & woods on chalk soils
CAREX ECHINATA	Star Sedge	Very rare; bogs on acid soils
Carex elata *	Tufted-sedge	1900 (Rush Green & Wolverton); ponds
CAREX FLACCA	Glaucous Sedge	Common; dry grassland & fens; often on calcareous soils
CAREX HIRTA	Hairy Sedge	Common; grassy places, often where damp
Carex hostiana *	Tawny Sedge	Pre-1926; marshes & wet meadows; was mostly recorded from the west
CAREX LAEVIGATA	Smooth-stalked Sedge	Very rare; marshy woodland on acid soils; *e.g.* Black Park
Carex montana *	Soft-leaved Sedge	1891 (Chalfont St Peter); woodland
CAREX MURICATA SSP. LAMPROCARPA	Small-fruited Prickly Sedge	Very rare; short turf on acid soils
CAREX MURICATA SSP. MURICATA	Large-fruited Prickly Sedge	NT; RDB; very rare; chalk grassland; recently discovered in three sites
CAREX NIGRA	Common Sedge	Rare; marshes, fens & wet meadows
CAREX OTRUBAE	False Fox-sedge	Common; wet meadows & by open water; rare south of The Chilterns
CAREX OVALIS	Oval Sedge	Scarce; meadows & heaths
CAREX PALLESCENS	Pale Sedge	Rare; woodland rides; often sporadic in appearance
CAREX PANICEA	Carnation Sedge	Scarce; wet grassland & fens on base-rich soils
CAREX PANICULATA	Greater Tussock-sedge	Scarce; marshes, fens, canal & river margins; decreasing
CAREX PENDULA	Pendulous Sedge	Locally frequent; woods, shady streamsides; also a garden escape; increasing
CAREX PILULIFERA	Pill Sedge	Scarce; woods & heaths on acid soils
CAREX PSEUDOCYPERUS	Cyperus Sedge	Rare; canal & pond margins; decreasing
CAREX PULICARIS	Flea Sedge	Very rare; wet acid grassland; now only known from Moorend Common
CAREX REMOTA	Remote Sedge	Frequent; damp woodland rides
CAREX RIPARIA	Greater Pond-sedge	Frequent; rivers, canals, ponds & marshy grassland

CAREX ROSTRATA	Bottle Sedge	Very rare; marshes
CAREX SPICATA	Spiked Sedge	Frequent; roadsides & damp grassland
CAREX STRIGOSA	Thin-spiked Wood-sedge	Scarce; damp woodland rides
CAREX SYLVATICA	Wood-sedge	Common; woodland rides & shady hedges
CAREX VESICARIA	Bladder-sedge	Very rare; marshy woodland on acid soils; *e.g.* Rowley Wood
CAREX VIRIDULA SSP. BRACHYRRHYNCHA	Long-stalked Yellow-sedge	Very rare; marshes in the north
CAREX VIRIDULA SSP. OEDOCARPA	Common Yellow-sedge	Rare; boggy grassland & woodland rides
CAREX VULPINA	True Fox-sedge	VU; RDB; Very rare; ditches in wet meadows; now only known from near Marsh Gibbon
Carex x csomadensis *		Pre-1926 (Grendon Underwood); meadows
Carex x fulva *		1904 (Lane End); bog
CAREX X PSEUDOAXILLARIS		Very rare; pond edges & woodland ride
Carex x sooi *		VU; pre-1926 (Longwick); bog
CARLINA VULGARIS	Carline Thistle	Scarce; dry calcareous grassland
CARPINUS BETULUS	Hornbeam	Locally common; woods, where it is probably native; planted elsewhere
Carthamus lanatus	Downy Safflower	1991 (Heelands); garden weed
Carthamus tinctoria	Safflower	Pre-1985 (Denham); rubbish heaps & waste places
Carum carvi	Caraway	EN; very rare; waste places; probably introduced as a seed or soil contaminant
CASTANEA SATIVA	Sweet Chestnut	Uncommon; woods on acid soils; always planted
CATABROSA AQUATICA	Whorl-grass	Scarce; ponds, streams & marshy grassland
Catalpa bignonioides	Indian Bean-tree	1999 (Little Missenden); planted in parkland & gardens
CATAPODIUM RIGIDUM	Fern-grass	Scarce; walls, dry bare ground on calcareous soils
CAUCALIS PLATYCARPOS *	Small Bur-parsley	EX; 1666 (Slough); cornfields
Cedrus atlantica	Atlas Cedar	Frequently planted in parks, large gardens & churchyards
Cedrus deodara	Deodar	Frequently planted in parks, large gardens & churchyards
Cedrus libani	Cedar-of-Lebanon	Frequently planted in parks, large gardens & churchyards
Centaurea calcitrapa *	Red Star-thistle	CR; 1961 (Aston Abbotts); waste places
CENTAUREA CYANUS	Cornflower	RDB; formerly frequent in cornfields, now only planted in wild flower seed mixes
Centaurea diluta *	Lesser Star-thistle	1903 (Iver); waste places
Centaurea melitensis *	Maltese Star-thistle	Pre-1926 (Taplow & Iver); rubbish heaps
Centaurea montana	Perennial Cornflower	Uncommon; waste ground
CENTAUREA NIGRA	Common Knapweed	Common; grassy places
CENTAUREA SCABIOSA	Greater Knapweed	Common; calcareous grassland
Centaurea solstitialis	Yellow Star-thistle	1994 (Haddenham); cornfields, waste places & gardens
CENTAURIUM ERYTHRAEA	Common Centaury	Frequent; dry open grassland & woodland rides
Centipeda minima	Spreading Sneezeweed	2005 (Woburn Sands); garden centre weed
Centranthus ruber	Red Valerian	Uncommon; walls, especially in urban areas
CEPHALANTHERA DAMASONIUM	White Helleborine	VU; scarce; beechwoods on The Chilterns (where not uncommon)
CEPHALANTHERA RUBRA	Red Helleborine	CR; RDB; very rare; one wood on The Chilterns
Cephalaria gigantea	Giant Scabious	Very rare; field at Fawley Bottom
CERASTIUM ARVENSE	Field Mouse-ear	1986 (Bacombe Warren); grassland
Cerastium diffusum	Sea Mouse-ear	Very rare; disused railway ballast at Bletchley
CERASTIUM FONTANUM SSP. HOLOSTEOIDES		1985 (Burnham Beeches); wet grassland
CERASTIUM FONTANUM SSP. VULGARE	Common Mouse-ear	Very common; grasslands, waste places & cultivated land
CERASTIUM GLOMERATUM	Sticky Mouse-ear	Common; grassy places, waste ground & short turf
CERASTIUM SEMIDECANDRUM	Little Mouse-ear	Very rare; short turf on acid soils
Cerastium tomentosum	Snow-in-summer	Uncommon; walls & banks
CERATOCAPNOS CLAVICULATA	Climbing Corydalis	Rare; woods & heaths on sandy soils; especially on the Brickhills
Ceratochloa carinata	California Brome	Rare; grassy places; often not persistant
Ceratochloa cathartica	Rescue Brome	1997 (Ilmer); farmyard
CERATOPHYLLUM DEMERSUM	Rigid Hornwort	Rare; rivers, canals & ponds; formerly more common
Ceratophyllum submersum *	Soft Hornwort	1904 (near Colnbrook); canals & ponds
Cercis siliquastrum	Judas Tree	1991 (Turville); churchyard
CETERACH OFFICINARUM	Rustyback	Scarce; old dry walls; increasing
Chaenomeles speciosa	Chinese Quince	1992 (Wolverton); abandoned garden & churchyards
CHAENORRHINUM MINUS	Small Toadflax	Uncommon; railway ballast, arable fields, rough ground & gardens
Chaerophyllum aureum *	Golden Chervil	1957 (Little Chalfont); grassy places

CHAEROPHYLLUM TEMULUM	Rough Chervil	Common; shady hedgerows & woodland edges
Chamaecyparis lawsoniana	Lawson's Cypress	Frequently planted in woods & gardens
Chamaecyparis nootkatensis	Nootka Cypress	Occasionally planted in woods & gardens
Chamaecyparis pisifera	Sawara Cypress	Occasionally planted in woods & gardens
Chamaemelum nobile *	Chamomile	VU; pre-1926; heathy places in the south
CHAMERION ANGUSTIFOLIUM	Rosebay Willowherb	Very common; roadsides, waste places & railway banks
CHELIDONIUM MAJUS	Greater Celandine	Uncommon; waste places near habitation
CHENOPODIUM ALBUM	Fat-hen	Common; roadsides, waste & cultivated ground
CHENOPODIUM BONUS-HENRICUS	Good-King-Henry	VU; rare; fields & waste places; much decreased
CHENOPODIUM FICIFOLIUM	Fig-leaved Goosefoot	Rare; disturbed & waste ground
Chenopodium glaucum *	Oak-leaved Goosefoot	VU; 1948 (Burnham Beeches); waste places
Chenopodium hybridum	Maple-leaved Goosefoot	Very rare; waste places e.g. Aylesbury
Chenopodium leptophyllum agg. *	Slimleaf Goosefoot	1904 (Taplow); waste ground & rubbish heaps
Chenopodium murale	Nettle-leaved Goosefoot	VU; 1996 (Haddenham); rubbish heaps
Chenopodium opulifolium *	Grey Goosefoot	Pre-1926; waste places in the south
CHENOPODIUM POLYSPERMUM	Many-seeded Goosefoot	Uncommon; arable fields & waste places
Chenopodium quinoa	Quinoa	Occasionally planted for pheasant food
CHENOPODIUM RUBRUM	Red Goosefoot	Uncommon; manure heaps & other rich soil
Chenopodium strictum *	Striped Goosefoot	1913 (Langley); waste ground
Chenopodium urbicum	Upright Goosefoot	CR; 1994 (Turweston); waste places & arable fields
Chenopodium vulvaria	Stinking Goosefoot	EN; RDB; 1981 (Burnham Beeches); waste places & farmyards
Chionodoxa forbesii	Glory-of-the-snow	Rare; parks & churchyards
Chionodoxa luciliae	Boisser's Glory-of-the-snow	Very rare; churchyard at Slough
Chionodoxa sardensis	Lesser Glory-of-the-snow	Very rare; churchyard at Slough
CHRYSANTHEMUM SEGETUM	Corn Marigold	VU; very rare; arable fields on sandy soils; much decreased
CHRYSOSPLENIUM OPPOSITIFOLIUM	Opposite-leaved Golden-saxifrage	Rare; streamsides & boggy areas in woods
Cicer arietinum	Chick Pea	1982 (Cippenham); waste places
Cicerbita bourgaei	Pontic Blue-sow-thistle	Very rare; long established by a roadside at Nether Winchendon
Cicerbita macrophylla ssp. uralensis	Blue-sow-thistle	Rare; roadsides close to habitation
CICHORIUM INTYBUS	Chicory	Locally common; roadsides & field margins; more frequent in the south
Cicuta virosa *	Cowbane	1778 (near Colnbrook); riverbanks
CIRCAEA LUTETIANA	Enchanter's-nightshade	Common; woods & gardens
CIRSIUM ACAULE	Dwarf Thistle	Uncommon; short calcareous turf
CIRSIUM ARVENSE	Creeping Thistle	Very common; rank grassland, waste places & roadsides
CIRSIUM DISSECTUM	Meadow Thistle	Very rare; wet grassland in mid-Bucks
CIRSIUM ERIOPHORUM	Woolly Thistle	Rare; calcareous grassland; especially in the north
CIRSIUM PALUSTRE	Marsh Thistle	Common; wet woodland rides, marshes, fens & other wet grassland
CIRSIUM VULGARE	Spear Thistle	Very common; rank grassland, waste places & roadsides
CIRSIUM X CELAKOVSKIANUM *		Pre-1926 (between Gerrards Cross & Chalfont); waste ground
Citrullus lanatus	Water Melon	1970 (Iver); rubbish dump
CLADIUM MARISCUS	Great Fen-sedge	1904 (between Wotton Underwood & Brill); pond; also rarely planted, as at Tongwell Lake
Claytonia perfoliata	Springbeauty	Rare; sandy places; locally abundant in the Brickhills
Claytonia sibirica	Pink Purslane	Rare; damp, shady places near habitation
Clematis montana	Himalayan Clematis	Very rare; garden escape, sometimes self-sown
CLEMATIS VITALBA	Traveller's-joy	Common; hedgerows & woods on calcareous soils
Clematis viticella	Purple Clematis	Very rare; grassland by River Ouzel at Caldecotte
CLINOPODIUM ACINOS	Basil Thyme	VU; rare; short calcareous turf in The Chilterns
CLINOPODIUM ASCENDENS	Common Calamint	Very rare; calcareous banks; now only in the south
CLINOPODIUM CALAMINTHA	Lesser Calamint	VU; NS; very rare; calcareous banks & walls
CLINOPODIUM VULGARE	Wild Basil	Common; hedges, wood borders & railway banks; especially on calcareous soils
Cochlearia danica	Danish Scurvygrass	Scarce; salted roadsides; increasing

COELOGLOSSUM VIRIDE	Frog Orchid	VU; very rare; chalk grassland; now only known from near Ivinghoe & Aston Clinton
Colchicum autumnale	Meadow Saffron	NT; 1984 (Bockmer End); waste ground
Colutea arborescens	Bladder-senna	Very rare; fields & hedges
Colutea x media	Orange Bladder-senna	1997 (Princes Risborough); hedge
CONIUM MACULATUM	Hemlock	Common; by rivers, canals, ditches & roadsides
CONOPODIUM MAJUS	Pignut	Uncommon; woods & fields; not on alkaline soils
Conringia orientalis *	Hare's-ear Mustard	c.1940 (Burcott); waste places
Consolida ajacis	Larkspur	Rare; waste places & roadsides
Consolida regalis *	Forking Larkspur	1861 (High Wycombe); roadside
CONVALLARIA MAJALIS	Lily-of-the-valley	Very rare; woods on acid soils; e.g. Black Park & Great Brickhill
CONVOLVULUS ARVENSIS	Field Bindweed	Very common; hedges, grassland, arable fields & waste places
Convolvulus tricolor	Tricolour Convolvulus	1970 (Beaconsfield); waste ground
Conyza bilbaoana	Bilbao Fleabane	1998 (High Wycombe); waste places
Conyza canadensis	Canadian Fleabane	Uncommon; urban weed; pavement cracks & waste places
Conyza sumatrensis	Guernsey Fleabane	Very rare; waste places & set-aside fields; increasing
Coreopsis tinctoria *	Garden Tickseed	1917 (Naphill); waste ground
Coriandrum sativum	Coriander	Very rare; waste places
Cornus alba	White Dogwood	Very rare; mass-planted in urban areas; rarely self-sown
Cornus mas	Cornelian-cherry	1986 (Old Rectory Wood); woodland
CORNUS SANGUINEA	Dogwood	Very common; hedges on calcareous soils; often invasive on chalk grassland
Cornus sericea	Red-osier Dogwood	Scarce; parkland & by ponds; sometimes well established by suckering
Coronopus didymus	Lesser Swine-cress	Rare; waste places & cultivated ground; less frequent in the north
CORONOPUS SQUAMATUS	Swine-cress	Common; paths & farmyard gateways
Corydalis solida	Bird-in-a-bush	Very rare; parkland e.g. Newport Pagnell
CORYLUS AVELLANA	Hazel	Very common; woodland & hedges
Corylus maximus	Filbert	Very rare; woodland; e.g. Bradenham Woods
Cosmos bipinnatus	Mexican Aster	Very rare; waste places near gardens
Cotinus coggygria	Smoke-tree	1999 (Heelands); waste places
Cotoneaster atropurpureus	Purple-flowered Cotoneaster	1983 (Newport Pagnell); churchyard
Cotoneaster bullatus	Hollyberry Cotoneaster	1996 (Heelands); pathside
Cotoneaster dielsianus	Diels' Cotoneaster	1998 (Newton Longville); pavement cracks
Cotoneaster divaricatus	Spreading Cotoneaster	1974 (Burnham Beeches); hedgerow
Cotoneaster frigidus	Tree Cotoneaster	1997 (New Bradwell); brookside
Cotoneaster hjelmqvistii	Hjelmqvist's Cotoneaster	1997 (Great Linford); woodland
Cotoneaster horizontalis	Wall Cotoneaster	Common; walls close to habitation; often self-sown
Cotoneaster integrifolius	Small-leaved Cotoneaster	1996 (Downley); roadsides
Cotoneaster lacteus	Late Cotoneaster	1993 (Hardwick); churchyard
Cotoneaster salicifolius	Willow-leaved Cotoneaster	1996 (Central Milton Keynes); seedlings in pavement cracks
Cotoneaster simonsii	Himalayan Cotoneaster	Uncommon; bird-sown in chalk grassland, woods & waste places
Cotoneaster x watereri	Waterer's Cotoneaster	1999 (Downs Barn); shrubberies & embankments
Cotula coronopifolia	Buttonweed	Very rare; pond & lake margins; e.g. Tongwell Lake
Crassula helmsii	New Zealand Pigmyweed	Scarce; ponds & lakes; increasing
Crataegus crus-galli	Cockspur-thorn	Occasional; planted in churchyards and parks; e.g. Bradenham
CRATAEGUS LAEVIGATA	Midland Hawthorn	Uncommon; woods & hedges; especially in the north
CRATAEGUS MONOGYNA	Hawthorn	Very common; hedges & woods
Crataegus persimilis	Broad-leaved Cockspur-thorn	1996 (Bradwell Common); hedge & pavement cracks
CRATAEGUS X MEDIA	Hybrid Hawthorn	Common; hedges & wood margins
CREPIS BIENNIS	Rough Hawk's-beard	Very rare; waste places; possibly overlooked
CREPIS CAPILLARIS	Smooth Hawk's-beard	Very common; short turf; often in churchyards
Crepis nicaeensis *	French Hawk's-beard	1897 (near Wendover); cultivated fields

Crepis setosa	Bristly Hawk's-beard	1989 (Salden); grassy & cultivated ground; introduced with grass seed
Crepis vesicaria ssp. taraxacifolia	Beaked Hawk's-beard	Common; roadsides & grassy places
Crocosmia paniculata	Aunt-Eliza	1994 (Ravenstone); roadside
Crocosmia x crocosmiiflora	Montbretia	Rare; waste places & roadsides
Crocus chrysanthus	Golden Crocus	Rare; churchyards & grassy places
Crocus chrysanthus x biflorus		1999 (Central Milton Keynes); waste ground
Crocus nudiflorus	Autumn Crocus	1992 (Hughenden); roadside
Crocus speciosus	Bieberstein's Crocus	1998 (Halton); churchyard
Crocus tommasinianus	Early Crocus	Rare; churchyards & waste ground
Crocus vernus ssp. vernus	Spring Crocus	Rare; churchyards & grassy places
Crocus vernus x tommasinianus		1998 (Swanbourne cemetery); short turf
Crocus x stellaris	Yellow Crocus	Rare; churchyards & grassy places
CRUCIATA LAEVIPES	Crosswort	Scarce; hedgerows & grassy places
Cryptomeria japonica	Japanese Red-cedar	1999 (Soulbury); parkland
Cucumis melo	Melon	Pre-1985 (Iver); rubbish heap
Cucumis sativus	Cucumber	Very rare; waste places & rubbish heaps
Cucurbita pepo	Marrow	2003 (Langley); waste places & canalsides
Cuphea ignea	Cigar Plant	2005 (Bletchley); garden weed
Cupressus macrocarpa	Monterey Cypress	Rare; churchyards & parkland
*Cuscuta epilinum**	Flax Dodder	1867 (Unlocalised); flaxfields
CUSCUTA EPITHYMUM	Dodder	VU; very rare; chalk grassland; heaths & clover fields
CUSCUTA EUROPAEA	Greater Dodder	NS; rare; riverbanks; usually on nettle; now only by the Rivers Great Ouse, Ouzel & Thames
Cyclamen hederifolium	Sowbread	Rare; hedges & churchyards
Cyclamen repandum	Spring Sowbread	1996 (Ellesborough); churchyard
Cymbalaria muralis ssp. muralis	Ivy-leaved Toadflax	Common; walls near habitation; very rarely on rubble heaps or canal locks
Cymbalaria pallida	Italian Toadflax	Very rare; wall at Bryants Bottom
Cynodon dactylon	Bermuda-grass	1970 (Chesham); waste places
Cynoglossum germanicum *	Green Hound's-tongue	Pre-1926 (near Mentmore); woods
CYNOGLOSSUM OFFICINALE	Hound's-tongue	NT; very rare; chalk grassland; *e.g.* Grangelands
CYNOSURUS CRISTATUS	Crested Dog's-tail	Very common; old grassland
Cynosurus echinatus	Rough Dog's-tail	1943 (Loudwater); waste places & gravel pits
Cyperus eragrostis	Pale Galingale	1994 (Bolter End); garden weed
CYPERUS FUSCUS	Brown Galingale	VU; RDB; very rare; seasonal pond & ditch at Dorney Common
Cyperus longus	Galingale	NT; NS; very rare; planted in ponds & by streams
Cystopteris fragilis	Brittle Bladder-fern	Very rare; walls; now only in Great Missenden churchyard
Cytisus battandieri	Pineapple Broom	1995 (Downs Barn); waste ground
Cytisus multiflorus	White Broom	1999 (near Richings Park); bank
CYTISUS SCOPARIUS SSP. SCOPARIUS	Broom	Scarce; heaths & roadsides on sandy soils
XCupressocyparis leylandii	Leyland Cypress	Scarce; woods, parkland & gardens
Daboecia cantabrica	St Dabeoc's Heath	c.1955 (Cliveden); parkland
DACTYLIS GLOMERATA	Cock's-foot	Very common; grassland
Dactylis polygama	Slender Cock's-foot	Very rare; woods at Cliveden
DACTYLORHIZA FUCHSII	Common Spotted-orchid	Locally common; chalk grassland, damp meadows & woods
DACTYLORHIZA INCARNATA SSP. INCARNATA	Early Marsh-orchid	Very rare; marshes & chalk grassland
Dactylorhiza incarnata ssp. pulchella		1972 (Hyde Lane); wet meadow
DACTYLORHIZA MACULATA SSP. ERICETORUM	Heath Spotted-orchid	Very rare; wet heaths; *e.g.* Moorend Common
DACTYLORHIZA PRAETERMISSA	Southern Marsh-orchid	Very rare; wet meadows
Dactylorhiza purpurella	Northern Marsh-orchid	Very rare; lakeside; introduced at College Lake
DACTYLORHIZA X GRANDIS		Very rare; marshes; now only at Moorend Common
Dactylorhiza x kerneriorum		1970 (Hyde Lane); wet meadow
Dahlia pinnata	Dahlia	1996 (Dibden Hill); rubbish heap
DAMASONIUM ALISMA	Starfruit	CR; RDB; very rare; ponds on acid soils in the south
DANTHONIA DECUMBENS	Heath-grass	Scarce; short acid grassland
DAPHNE LAUREOLA	Spurge-laurel	Uncommon; hedges & beechwoods
DAPHNE MEZEREUM	Mezereon	VU; NS; very rare; woods & scrub; now only known at two sites in The Chilterns
Darmera peltata	Indian-rhubarb	2001 (Biddlesden); wet woodland & parkland

Datura metel *	Devil's Trumpet	Pre-1970 (Aston Abbotts); mangel field
Datura stramonium	Thorn-apple	Very rare; roadsides & waste places
DAUCUS CAROTA SSP. CAROTA	Wild Carrot	Common; calcareous grassland & roadsides
Daucus carota ssp. sativus	Carrot	1986 (Turweston); waste places
DESCHAMPSIA CESPITOSA SSP. CESPITOSA	Tufted Hair-grass	Common; damp grassland & fens
DESCHAMPSIA CESPITOSA SSP. PARVIFLORA	Small-flowered Hair-grass	Uncommon; woods; probably under-recorded
Deschampsia danthonioides	Annual Hair-grass	1978 (Little Brickhill); golf course
DESCHAMPSIA FLEXUOSA	Wavy Hair-grass	Uncommon; woods on acid soils
Descurainia sophia	Flixweed	Very rare; waste places & cultivated fields; sporadic
Dianthus armeria *	Deptford Pink	EN; 1917 (Chalfont St Peter); sandy fields & hedgebanks
Dianthus barbatus	Sweet-William	2003 (Lane End); waste places
DIANTHUS DELTOIDES	Maiden Pink	NS; 1805 (Salt Hill); dry places; now occurs rarely as a garden escape
Dianthus gratianopolitanus x plumaris		1998 (Aylesbury); roadside
Dicentra spectabilis	Asian Bleeding-heart	2003 (Moor Wood); woods
DIGITALIS PURPUREA	Foxglove	Locally common; woods & open places on acid soils; also a garden escape
Digitalis x fucata		1997 (Haddenham); spontaneous in a garden
Digitaria ischaemum *	Smooth Finger-grass	Pre-1926 (Taplow); waste places
Digitaria sanguinalis	Hairy Finger-grass	1982 (Stewkley); gardens & waste places
Diplotaxis muralis	Annual Wall-rocket	Very rare; waste places & walls; especially in the south
DIPLOTAXIS TENUIFOLIA	Perennial Wall-rocket	Very rare; roadsides & waste places in the south
DIPSACUS FULLONUM	Wild Teasel	Common; damp grassland & churchyards
DIPSACUS PILOSUS	Small Teasel	Very rare; hedges, woods & by water
Dipsacus sativus *	Fuller's Teasel	Pre-1926 (Langley); waste places
Doronicum pardalianches	Leopard's-bane	Very rare; woods & parkland
Doronicum plantagineum	Plantain-leaved Leopard's-bane	1985 (Hedsor); roadside
Doronicum x excelsum	Harpur-Crewe's Leopard's-bane	1999 (Woburn Sands); woodland
Dorycnium hirsutum	Canary Clover	1993 (Bradenham); churchyard
Downingia elegans	Californian Lobelia	1989 (Salden); grassland; introduced with imported seed
Draba muralis	Wall Whitlow-grass	Very rare; walls & gravel paths; *e.g.* Wycombe Abbey
Dracocephalum thymiflorum *	Thyme-leaved Dragonhead	1871 (near High Wycombe); clover field
Dracunculus vulgaris	Dragon Arum	1995 (Hyde Lane); plantation
DROSERA INTERMEDIA	Oblong-leaved Sundew	1993 (Stoke Common); wet heaths
Drosera rotundifolia	Round-leaved Sundew	1980 (Burnham Beeches); bogs
DRYOPTERIS AFFINIS SSP. AFFINIS	Western Scaly Male-fern	Very rare; woods; *e.g.* Grubbins Plantation
DRYOPTERIS AFFINIS SSP. BORRERI	Common Scaly Male-fern	Rare; woods
DRYOPTERIS CARTHUSIANA	Narrow Buckler-fern	Scarce; woods
DRYOPTERIS DILATATA	Broad Buckler-fern	Locally common; woods
DRYOPTERIS FILIX-MAS	Male-fern	Common; woods, hedges, walls & churchyards; sometimes a garden escape
DRYOPTERIS X DEWEVERI		Very rare; woodland at Burnham Beeches
Duchesnea indica	Yellow-flowered Strawberry	Very rare; grassy places & paths; *e.g.* Wycombe Abbey
XDactyloglossum mixtum *		1969 (Pink Hill); chalk grassland
Echinochloa colona	Shama Millet	1970 (Iver); rubbish heap
Echinochloa crus-galli	Cockspur	Rare; arable fields; also a casual in waste places
Echinochloa frumentacea	White Millet	Pre-1985 (Denham); waste places
Echinops bannaticus	Blue Globe-thistle	1997 (Princes Risborough); roadsides & waste places
Echinops exaltatus	Globe-thistle	1997 (Little Missenden); pathside & waste places
Echinops sphaerocephalus	Glandular Globe-thistle	1983 (Chesham); waste ground
Echium plantagineum	Purple Viper's-bugloss	2003 (Moor Wood); wood & roadside
ECHIUM VULGARE	Viper's-bugloss	Very rare; dry sunny banks on The Chilterns
Elaeagnus umbellata	Spreading Oleaster	1997 (Heelands); waste ground
Eleocharis acicularis	Needle Spike-rush	1972 (Stoke Common); canals, rivers & ponds
ELEOCHARIS MULTICAULIS	Many-stalked Spike-rush	Very rare; bog; now only known from Burnham Beeches

ELEOCHARIS PALUSTRIS SSP. VULGARIS	Common Spike-rush	Scarce; pond & canal edges
Eleocharis quinqueflora *	Few-flowered Spike-rush	Pre-1926 (Dropmore, Cippenham & Chesham Moor); marshes
ELEOGITON FLUITANS	Floating Club-rush	Very rare; seasonal ponds; e.g. Stoke Common
Elodea canadensis	Canadian Waterweed	Uncommon; ponds & lakes; decreasing
Elodea nuttallii	Nuttall's Waterweed	Scarce; ponds, lakes & streams; increasing
ELYMUS CANINUS	Bearded Couch	Uncommon; woodland edges & old hedgerows
Elymus virginicus *	Virginia Wild-rye	Pre-1960 (unlocalised); waste places
ELYTRIGIA REPENS SSP. REPENS	Common Couch	Very common; rough grassland
Enarthrocarpus lyratus *		Pre-1926 (Eton & Iver); waste places
Epilobium brunnescens	New Zealand Willowherb	1996 (Heelands); garden weed
Epilobium ciliatum	American Willowherb	Common; waste places, woods & churchyards
EPILOBIUM HIRSUTUM	Great Willowherb	Very common; by water & in damp grassland
EPILOBIUM LANCEOLATUM	Spear-leaved Willowherb	1993 (Beaconsfield); roadsides
Epilobium lanceolatum x ciliatum *		1961 (Bourne End)
EPILOBIUM MONTANUM	Broad-leaved Willowherb	Common; waste places, woods & churchyards
EPILOBIUM OBSCURUM	Short-fruited Willowherb	Frequent; waste places & churchyards
EPILOBIUM PALUSTRE	Marsh Willowherb	Very rare; fens; decreasing
EPILOBIUM PARVIFLORUM	Hoary Willowherb	Uncommon; damp grassland & waste places
EPILOBIUM ROSEUM	Pale Willowherb	Rare; waste places, ditches & gardens
EPILOBIUM TETRAGONUM	Square-stalked Willowherb	Common; waste places, woods & churchyards
Epilobium x abortivum *		1969 (Bourne End)
Epilobium x aggregatum *		1958 (Burnham Beeches)
Epilobium x borbasianum *		1955 (Cores End)
Epilobium x brachiatum *		1954 (Bourne End)
Epilobium x dacicum *		1953 (Burnham Beeches)
Epilobium x erroneum *		1958 (Bourne End)
Epilobium x fallacinum *		1960 (Bourne End)
Epilobium x floridulum *		1958 (Burnham Beeches)
Epilobium x haussknechtianum *		1964 (Bourne End)
Epilobium x interjectum		Rare; waste places, woods & churchyards
Epilobium x limosum *		1957 (Bourne End)
Epilobium x mentiens *		1956 (Bourne End)
Epilobium x mutabile *		1956 (Bourne End)
Epilobium x neogradense *		1972 (Bourne End)
Epilobium x novae-civitatis *		1953 (Burnham Beeches)
Epilobium x nutantiflorum *		1963 (Bourne End)
Epilobium x palatinum *		1956 (Bourne End)
Epilobium x persicinum *		1955 (Bourne End)
Epilobium x schmidtianum *		Pre-1937 (unlocalised)
Epilobium x subhirsutum *		1953 (Burnham Beeches)
Epilobium x vicinum *		1959 (Bourne End)
Epimedium alpinum *	Barren-wort	1875 (Calverton Park); parkland
EPIPACTIS HELLEBORINE	Broad-leaved Helleborine	Scarce; woods; also rarely in chalk grassland
EPIPACTIS MUELLERI	Narrow-lipped Helleborine	NS; rare; woods on The Chilterns
EPIPACTIS PALUSTRIS	Marsh Helleborine	Very rare; fens
EPIPACTIS PHYLLANTHES	Green-flowered Helleborine	1989 (Black Park); woodland
EPIPACTIS PURPURATA	Violet Helleborine	Rare; open areas in woods on The Chilterns
EPIPOGIUM APHYLLUM	Ghost Orchid	EX; RDB; 1987; open woodland on The Chilterns
EQUISETUM ARVENSE	Field Horsetail	Common; rough grassland & gardens
EQUISETUM FLUVIATILE	Water Horsetail	Rare; ponds & lakes
EQUISETUM PALUSTRE	Marsh Horsetail	Uncommon; wet grassland & fens
EQUISETUM SYLVATICUM	Wood Horsetail	Very rare; woods on acid soils in the south
EQUISETUM TELMATEIA	Great Horsetail	Scarce; wet grass, woods & rough areas on less alkaline soils
EQUISETUM X LITORALE	Shore Horsetail	1994 (Linford Pits); lake margin
Eranthis hyemalis	Winter Aconite	Uncommon; woods, parkland & churchyards
ERICA CINEREA	Bell Heather	Very rare; dry heaths in the south
ERICA TETRALIX	Cross-leaved Heath	Very rare; wet heaths in the south
ERIGERON ACER	Blue Fleabane	Scarce; dry, bare places
Erigeron annuus	Tall Fleabane	Very rare; waste ground at Wycombe Marsh
Erigeron karvinskianus	Mexican Fleabane	Rare; walls & pavement cracks

Erinus alpinus	Fairy Foxglove	Very rare; walls; *e.g.* Princes Risborough
ERIOPHORUM ANGUSTIFOLIUM	Common Cotton-grass	Very rare; wet grassland & fens
ERODIUM CICUTARIUM	Common Stork's-bill	Rare; dry sandy places & road verges
Erodium glaucophyllum *	Glaucous Stork's-bill	1958 (Wooburn); waste places
Erodium moschatum	Musk Stork's-bill	Pre-1926 (Salt Hill & Ditton); waste places
EROPHILA MAJUSCULA	Hairy Whitlowgrass	1999 (Rushmere Park); short turf on sandy soil
EROPHILA VERNA	Common Whitlowgrass	Common; bare soil, walls, roadsides
Eruca vesicaria ssp. sativa	Garden Rocket	1994 (Aylesbury); grassy & waste places
Erucaria hispanica	Spanish Mustard	Pre-1926 (Princes Risborough); waste ground
Eryngium campestre	Field Eryngo	CR; 2003 (Central Milton Keynes); road verge
Eryngium planum	Blue Eryngo	2004 (Woughton Park); roadsides
Eryngium tripartitum	Thorn-bush Thistle	Pre-1985 (Slough); waste places
ERYSIMUM CHEIRANTHOIDES	Treacle-mustard	Scarce; arable fields
ERYSIMUM CHEIRI	Wallflower	Uncommon; walls; especially on limestone
Erysimum repandum *	Spreading Treacle-mustard	Pre-1926 (Iver); waste places
Erysimum x marshallii	Siberian Wallflower	1995 (Bradville); waste ground
Escallonia macrantha	Escallonia	1998 (Middle Green); canalside
Eschscholzia californica	Californian Poppy	Rare; waste places
Eucalyptus gunnii	Cider Gum	Very rare; woods; planted; *e.g.* Linford Wood
Euclidium syriacum *	Syrian Mustard	Pre-1926 (Iver); waste places
EUONYMUS EUROPAEUS	Spindle	Uncommon; hedges & woodland margins on calcareous soils
Euonymus fortunei	Wintercreeper	1996 (Heelands); waste places
Euonymus japonicus	Evergreen Spindle	Rare; hedges & churchyards; probably always planted
EUPATORIUM CANNABINUM	Hemp-agrimony	Uncommon; pond & canal margins, wet woodland rides
EUPHORBIA AMYGDALOIDES SSP. AMYGDALOIDES	Wood Spurge	Locally common; woods; especially from The Chilterns southwestwards
Euphorbia amygdaloides ssp. robbiae	Mrs Robb's Bonnet	Uncommon; hedges; *e.g.* Bradwell Common
Euphorbia characias	Mediterranean Spurge	1997 (Central Milton Keynes); waste ground
Euphorbia cyparissias	Cypress Spurge	Rare; grassy & waste places
Euphorbia dulcis	Sweet Spurge	Very rare; hedges & plantations
EUPHORBIA EXIGUA	Dwarf Spurge	NT; uncommon; arable fields; especially on calcareous soils
EUPHORBIA HELIOSCOPIA	Sun Spurge	Common; gardens & fallow fields
Euphorbia lathyris	Caper Spurge	Uncommon; waste ground & gardens
Euphorbia myrsinites	Blue Spurge	1994 (near Ravenstone); on the site of a demolished house
Euphorbia oblongata	Balkan Spurge	1998 (Shabbington); churchyard
EUPHORBIA PEPLUS	Petty Spurge	Common; gardens, churchyards & waste places
EUPHORBIA PLATYPHYLLOS	Broad-leaved Spurge	NS; rare; arable fields & roadsides; sporadic in appearance
Euphorbia serrulata	Upright Spurge	1999 (Hughenden Valley); gardens & old allotments
Euphorbia waldsteinii	Waldstein's Spurge	1998 (Amerden); waste ground near water
Euphorbia x pseudovirgata	Twiggy Spurge	2000 (Princes Risborough); railway bank
Euphrasia anglica	English Sticky Eyebright	EN; pre-1985 (near Cliveden); short turf
Euphrasia confusa *	Little Kneeling Eyebright	Pre-1970 (unlocalised)
Euphrasia micrantha *	Common Slender Eyebright	Pre-1926 (Naphill Common & near Amersham); heathy places
EUPHRASIA NEMOROSA	Common Eyebright	Scarce; dry grassy places on poor soils
EUPHRASIA PSEUDOKERNERI	Chalk-hill Eyebright	EN; NS; rare; chalk grassland
Euphrasia rostkoviana *	Mountain Sticky Eyebright	VU; pre-1926 (Penn Street, Cholesbury & Chesham); damp grassland
Euphrasia tetraquetra *	Broad-leaved Eyebright	1923 (Ivinghoe Beacon); chalk grassland
Fagopyrum esculentum	Buckwheat	Rare; fields & waste places; usually introduced with pheasant food
Fagopyrum tataricum *	Green Buckwheat	Pre-1926 (Langley); waste places
FAGUS SYLVATICA	Beech	Common; woods; especially on The Chilterns
Falcaria vulgaris	Longleaf	Very rare; grassy banks; now only known from Cryers Hill
Fallopia baldschuanica	Russian-vine	Rare; over hedges & trees
FALLOPIA CONVOLVULUS	Black-bindweed	Common; arable fields, gardens & waste places
FALLOPIA DUMETORUM	Copse-bindweed	VU; NS; 1973 (High Wycombe); hedges
Fallopia japonica	Japanese Knotweed	Uncommon; roadsides, waste places & gardens
Fallopia sachalinensis	Giant Knotweed	Very rare; roadsides, churchyards & waste places

Fallopia x bohemica	Hybrid Knotweed	Very rare; only known from Spade Oak & Black Park
Fargesia murieliae	Umbrella Bamboo	2002 (High Wycombe); riverbank; died after flowering
Festuca altissima	Wood Fescue	1970 (Cliveden); woodland
FESTUCA ARUNDINACEA	Tall Fescue	Common; rank grassland & roadsides; especially on damp clay
Festuca brevipila	Hard Fescue	1997 (Heelands); road verge; introduced with grass seed
FESTUCA FILIFORMIS	Fine-leaved Sheep's-fescue	1998 (Burnham Beeches); sandy grassland; probably overlooked
FESTUCA GIGANTEA	Giant Fescue	Common; woods & shady hedgerows; especially on clay soils
Festuca heterophylla	Various-leaved Fescue	1985 (Cliveden); woodland
FESTUCA OVINA	Sheep's-fescue	Uncommon; short turf on acid or alkaline soils; under-recorded
FESTUCA PRATENSIS	Meadow Fescue	Common; old meadows
Festuca rubra ssp. commutata	Chewing's Fescue	1997 (Denham); grassland
Festuca rubra ssp. megastachys		1992 (Bletchley); grassland
FESTUCA RUBRA SSP. RUBRA	Red Fescue	Very common; grassland, roadsides & walls
Ficus carica	Fig	1984 (Marlow); churchyard
FILAGO GALLICA *	Narrow-leaved Cudweed	EX; c.1774 (near Iver); dry sandy fields
Filago lutescens *	Red-tipped Cudweed	EN; RDB; 1898 (Great Brickhill); dry sandy fields
FILAGO MINIMA	Small Cudweed	Very rare; sandy heaths; e.g. Old Wavendon Heath
FILAGO PYRAMIDATA *	Broad-leaved Cudweed	EN; RDB; pre-1926; dry fields & roadsides in the south
FILAGO VULGARIS	Common Cudweed	NT; very rare; dry sandy fields; much decreased
FILIPENDULA ULMARIA	Meadowsweet	Common; wet grasslands, damp woods & by water
FILIPENDULA VULGARIS	Dropwort	Rare; dry calcareous grassland & wet meadows
FOENICULUM VULGARE	Fennel	Rare; waste places
Forsythia x intermedia	Forsythia	Very rare; rubbish heaps; much planted in parks & gardens
Fragaria moschata *	Hautbois Strawberry	Pre-1926 (Handy Cross & Forty Green); hedgebanks
FRAGARIA VESCA	Wild Strawberry	Uncommon; woods & hedgebanks; especially on calcareous soils
Fragaria x ananassa	Garden Strawberry	Rare; railwaybanks & waste places
FRANGULA ALNUS	Alder Buckthorn	Very rare; damp, acid woods in the south; planted elsewhere
FRAXINUS EXCELSIOR	Ash	Very common; woods & hedges
Fraxinus ornus	Manna Ash	Very rare; planted in parkland at Great Missenden
Fritillaria imperialis	Crown Imperial	1998 (Wendover); waste places
FRITILLARIA MELEAGRIS	Fritillary	VU; NS; very rare; wet meadows; much decreased; now only in the Thame valley
Fumaria agraria *	Field Fumitory	1903 (between Iver & Drayton); rubbish heaps
Fumaria capreolata ssp. babingtonii *	White Ramping-fumitory	Pre-1926 (Turville & Eton); cultivated ground
FUMARIA DENSIFLORA	Dense-flowered Fumitory	NS; rare; arable fields; much decreased
FUMARIA MURALIS SSP. BORAEI	Common Ramping-fumitory	1996 (Walter's Ash); arable fields & railwaybank
FUMARIA OFFICINALIS SSP. OFFICINALIS	Common Fumitory	Common; arable fields & waste places; especially on light soils
FUMARIA OFFICINALIS SSP. WIRTGENII		Pre-1985 (Gerrards Cross); arable fields
FUMARIA PARVIFLORA	Fine-leaved Fumitory	VU; NS; very rare; arable fields on chalk soils
Fumaria purpurea *	Purple Ramping-fumitory	NS; 1906 (near Leighton Buzzard); cultivated ground
FUMARIA VAILLANTII	Few-flowered Fumitory	VU; NS; rare; arable fields on chalk soils
XFESTULOLIUM BRAUNII		1994 (Whaddon); grassland
XFESTULOLIUM LOLIACEUM		Scarce; grassland; often on damp soils
Gagea lutea *	Yellow Star-of-Bethlehem	c.1950 (Great Brickhill); wet woodland
Galanthus elwesii	Greater Snowdrop	Rare; churchyards & pathsides; e.g. Bradwell Common
Galanthus nivalis	Snowdrop	Common; churchyards, roadsides & woods
Galanthus nivalis x plicatus		1995 (Ickford); churchyard
Galanthus woronowii	Green Snowdrop	Rare; churchyards & pathsides; e.g. Bradwell Common
Galega officinalis	Goat's-rue	Scarce; roadsides & waste places in the south
GALEOPSIS ANGUSTIFOLIA	Red Hemp-nettle	CR; NS; very rare; arable fields & railways
GALEOPSIS BIFIDA	Bifid Hemp-nettle	Rare; arable fields & waste places; probably under-recorded
GALEOPSIS SPECIOSA	Large-flowered Hemp-nettle	VU; 1984 (Pens Place); waste places
GALEOPSIS TETRAHIT	Common Hemp-nettle	Uncommon; woods, waysides & cultivated land

Galinsoga parviflora	Gallant-soldier	Rare; urban waste places & gardens; decreasing
Galinsoga quadriradiata	Shaggy-soldier	Rare; urban waste places & gardens; decreasing
GALIUM APARINE	Cleavers	Very common; woods, hedges, waste places & gardens
GALIUM MOLLUGO SSP. ERECTUM	Upright Bedstraw	Local; dry calcareous grassland
GALIUM MOLLUGO SSP. MOLLUGO	Hedge Bedstraw	Common; woods, hedges & grassy banks
GALIUM ODORATUM	Woodruff	Locally common; woods; especially on The Chilterns
GALIUM PALUSTRE SSP. ELONGATUM	Great Marsh-bedstraw	Rare; riverbanks; *e.g.* Stanton Low; probably under-recorded
GALIUM PALUSTRE SSP. PALUSTRE	Common Marsh-bedstraw	Common; marshes & by water
Galium parisiense*	Wall Bedstraw	NS; 1933 (between Eton & Eton Wick)
Galium pumilum*	Slender Bedstraw	EN; NS; 1897 (Aston Farm); chalk grassland
GALIUM SAXATILE	Heath Bedstraw	Uncommon; short, acid grassland
Galium spurium*	False Cleavers	Pre-1926 (Langley); brickyard
GALIUM TRICORNUTUM	Corn Cleavers	CR; RDB; 1971 (Downley); cornfields
GALIUM ULIGINOSUM	Fen Bedstraw	Scarce; fens & pond margins; especially in the north
GALIUM VERUM	Lady's Bedstraw	Common; short, mostly dry, grassland
Galium x pomeranicum		Pre-1985 (Shenley Church End); grassy places
Gastridium ventricosum *	Nit-grass	1943 (Wing); waste ground
Gaudinia fragilis	French Oat-grass	Pre-1985 (Pitstone)
Gaultheria procumbens *	Checkerberry	1957 (Langley Park); heathy places
Gaultheria shallon	Shallon	Very rare; heathy woodland; *e.g.* Littleworth Common
GENISTA ANGLICA	Petty Whin	NT; very rare; damp heaths; *e.g.* Moorend Common
GENISTA TINCTORIA SSP. TINCTORIA	Dyer's Greenweed	Very rare; damp grassland; now only known from the Ray valley
GENTIANELLA AMARELLA SSP. AMARELLA	Autumn Gentian	Rare; chalk downland; very rare off The Chilterns
GENTIANELLA ANGLICA	Early Gentian	NS; very rare; chalk downland; *e.g.* Pitstone Hills
GENTIANELLA CILIATA	Fringed Gentian	CR; RDB; very rare; still on a hill not far from Wendover
GENTIANELLA GERMANICA	Chiltern Gentian	NS; rare; open, dry chalk grassland
GENTIANELLA X PAMPLINII		Very rare; chalk downland; *e.g.* Whiteleaf Hill
GERANIUM COLUMBINUM	Long-stalked Crane's-bill	Rare; chalk grassland
GERANIUM DISSECTUM	Cut-leaved Crane's-bill	Very common; arable fields & waste places
Geranium endressii	French Crane's-bill	1991 (Hale Wood); road verge
GERANIUM LUCIDUM	Shining Crane's-bill	Scarce; walls, churchyards; hedgebanks
Geranium macrorrhizum	Rock Crane's-bill	Very rare; hedge & road verge; *e.g.* Chesham Moor
GERANIUM MOLLE	Dove's-foot Crane's-bill	Common; grassy places
Geranium nodosum	Knotted Crane's-bill	1986 (Hyde Heath); commonland
Geranium phaeum	Dusky Crane's-bill	1998 (Wendover); hedges
GERANIUM PRATENSE	Meadow Crane's-bill	Uncommon; roadsides & old meadows
GERANIUM PUSILLUM	Small-flowered Crane's-bill	Scarce; short grassland; not on calcareous soils
Geranium pyrenaicum	Hedgerow Crane's-bill	Uncommon; roadsides & hedgebanks
GERANIUM ROBERTIANUM SSP. ROBERTIANUM	Herb-Robert	Very common; woods, hedges & churchyards
GERANIUM ROTUNDIFOLIUM	Round-leaved Crane's-bill	Rare; hedges & urban areas
Geranium sanguineum	Bloody Crane's-bill	1995 (Naphill Common); waste places
Geranium versicolor	Pencilled Crane's-bill	1980 (Oakley Wood); woodland
Geranium x oxonianum	Druce's Crane's-bill	Rare; waste places
Geum rivale	Water Avens	1971 (Latimer); wet meadows
GEUM URBANUM	Wood Avens	Common; woods, hedges & churchyards
Geum x intermedium *	Hybrid Avens	1903 (near Stewkley); wet meadows
Ginkgo biloba	Maidenhair-tree	1996 (Milton Keynes Village); churchyard
Gladiolus communis ssp. byzantinus	Eastern Gladiolus	1999 (Halton); canalside
Glaucium corniculatum *	Red Horned-poppy	Pre-1926 (Iver); waste places
GLECHOMA HEDERACEA	Ground-ivy	Very common; woods, hedges, waste places & gardens
Globularia punctata	Common Globularia	Very rare; Stonepit Field, Great Linford
GLYCERIA DECLINATA	Small Sweet-grass	Rare; wet grassland & by ponds & streams
GLYCERIA FLUITANS	Floating Sweet-grass	Common; marshes, fens & by water
Glyceria fluitans x declinata		1975 (Howe Park Wood); woodland ride
GLYCERIA MAXIMA	Reed Sweet-grass	Common; riversides, ditches, lakes & ponds
GLYCERIA NOTATA	Plicate Sweet-grass	Uncommon; watersides
GLYCERIA X PEDICELLATA		Rare; watersides
Glycine max	Soya-bean	1970 (Weedon); roadside

GNAPHALIUM SYLVATICUM	Heath Cudweed	EN; 1981 (Rowley Wood); heathy woods
GNAPHALIUM ULIGINOSUM	Marsh Cudweed	Uncommon; muddy tracks in fields & waysides
Grindelia squarrosa *	Curly-cup Gumweed	1925 (Beaconsfield); waste places
GROENLANDIA DENSA	Opposite-leaved Pondweed	VU; very rare; streams in the south
Guizotia abyssinica	Niger	2000 (Soulbury); garden weed; introduced with birdseed
Gunnera manicata	Brazilian Giant-rhubarb	1988 (Dorton); pondside
Gunnera tinctoria	Giant-rhubarb	1988 (Cublington); pondside
GYMNADENIA CONOPSEA SSP. CONOPSEA	Chalk Fragrant-orchid	Rare; chalk downland; where it may be abundant
Gymnadenia conopsea ssp. densiflora *	Marsh Fragrant-orchid	Pre-1926 (Winslow); marshes
Gymnocarpium dryopteris *	Oak Fern	1883 (Ballard's Wood); woodland
GYMNOCARPIUM ROBERTIANUM	Limestone Fern	NS; 1996 (Wooburn); walls & woodland
Gypsophila paniculata *	Baby's Breath	Pre-1926 (Slough); waste places
Hedera algeriensis	Algerian Ivy	1997 (Little Missenden); roadside
Hedera colchica	Persian Ivy	1996 (Whelpley Hill); hedge
HEDERA HELIX SSP. HELIX	Common Ivy	Very common; woods, hedgerows, walls & gardens
Hedera helix ssp. hibernica	Atlantic Ivy	1986 (Peterley Wood); woodland
HELIANTHEMUM NUMMULARIUM	Common Rock-rose	Uncommon; open grassland on the chalk
Helianthus annuus	Sunflower	Uncommon; waste places
Helianthus petiolaris	Lesser Sunflower	2003 (Drayton Parslow); waste places
Helianthus tuberosus *	Jerusalem Artichoke	Pre-1926; waste places
Helianthus x laetiflorus *	Perennial Sunflower	1996 (Middle Claydon); waste places
Helichrysum bracteatum	Strawflower	1970 (Iver); rubbish heap
HELICTOTRICHON PRATENSE	Meadow Oat-grass	Scarce; old grassland on calcareous soils
HELICTOTRICHON PUBESCENS	Hairy Oat-grass	Uncommon; old meadows, pastures & churchyards
Helleborus dumetorum	Shrubby Hellebore	1980 (Boswells); copse
HELLEBORUS FOETIDUS	Stinking Hellebore	NS; very rare; woods & hedgebanks; also a garden escape
Helleborus niger	Christmas-rose	1982 (Ibstone); churchyard
Helleborus orientalis	Lenten-rose	Very rare; shaded road verge at Cryers Hill
HELLEBORUS VIRIDIS SSP. OCCIDENTALIS	Green Hellebore	Rare; woodland
Hemerocallis fulva	Orange Day-lily	1988 (Hedgerley Green); old gravel pit
Hemerocallis lilioasphodelus	Yellow Day-lily	1998 (Halton); pondside
Heracleum mantegazzianum	Giant Hogweed	Rare; roadsides & waste places
HERACLEUM SPHONDYLIUM SSP. SPHONDYLIUM	Hogweed	Very common; woods; hedges; roadsides & waste places
HERMINIUM MONORCHIS	Musk Orchid	VU; NS; very rare; open, short grassland on the chalk
Herniaria glabra	Smooth Rupturewort	2001 (Booker); garden centre weed
Hesperis matronalis	Dame's-violet	Scarce; roadsides near habitation
Heuchera sanguinea	Coralbells	1995 (Walk Wood); woodland
Hibiscus trionum *	Bladder Ketmia	Pre-1926 (Iver); waste places
Hieracium anglorum *		1957 (Burnham Beeches)
Hieracium argillaceum *		1968 (Dropmore)
Hieracium calcaricola		1978 (Fulmer)
Hieracium consociatum *		1962 (Princes Risborough)
Hieracium glevense *		1955 (Chorleywood)
Hieracium grandidens		1997 (Whiteleaf Hill)
Hieracium rigens *		1948 (Davenport Wood)
Hieracium sabaudum	Savoy Hawkweed	2000 (Bletchley)
HIERACIUM SALTICOLA		1996 (Moor End)
Hieracium scotostictum		1992 (Lucas Wood)
Hieracium spilophaeum	Spotted Hawkweed	1991 (Cliveden)
Hieracium trichocaulon		1984 (Stoke Common)
HIERACIUM UMBELLATUM	Leafy Hawkweed	*e.g.* Bow Brickhill
Hieracium vagum		1998 (High Wycombe)
HIERACIUM VIRGULTORUM		1990 (Beaconsfield)
Himantoglossum hircinum *	Lizard Orchid	NT; RDB; 1931 (Loudwater); chalk grassland
HIPPOCREPIS COMOSA	Horseshoe Vetch	Rare; sunny banks on calcareous soils
Hippophae rhamnoides	Sea-buckthorn	NS; very rare; waste ground; *e.g.* Calvert
HIPPURIS VULGARIS	Mare's-tail	Rare; ponds & rivers; especially in the south
Hirschfeldia incana	Hoary Mustard	Rare; waste places in the south; *e.g.* Slough arm of the Grand Union Canal
HOLCUS LANATUS	Yorkshire-fog	Very common; grassland

HOLCUS MOLLIS	Creeping Soft-grass	Common; heathy places & woods on acid soils
HORDELYMUS EUROPAEUS	Wood Barley	NS; locally common; woods on The Chilterns
Hordeum distichon	Two-rowed Barley	Frequent; roadsides, arable fields & waste places
Hordeum jubatum	Foxtail Barley	Rare; roadsides & waste places
HORDEUM MURINUM SSP. MURINUM	Wall Barley	Common; roadsides, waste places & bases of walls
HORDEUM SECALINUM	Meadow Barley	Common; old grassland
Hordeum vulgare	Six-rowed Barley	Uncommon; roadsides, arable fields & waste places
HOTTONIA PALUSTRIS	Water Violet	Very rare; ponds & ditches; *e.g.* Stowe Park
HUMULUS LUPULUS	Hop	Uncommon; hedgerows
HUPERZIA SELAGO SSP. SELAGO	Fir Clubmoss	1977 (Lodge Wood); heathy places
Hyacinthoides hispanica	Spanish Bluebell	Uncommon; churchyards & waste places
HYACINTHOIDES NON-SCRIPTA	Bluebell	Common; woods & old hedges
Hyacinthoides x massartiana	Hybrid Bluebell	Uncommon; churchyards & waste places
Hyacinthus orientalis	Hyacinth	1996 (unlocalised); waste places
Hydrangea macrophylla	Hydrangea	2003 (Park Wood); woodland
Hydrocharis morsus-ranae	Frogbit	VU; 1982 (Temple Island Meadows); ponds & ditches
Hydrocotyle ranunculoides	Floating Pennywort	Very rare; ponds & lakes; *e.g.* Soulbury
HYDROCOTYLE VULGARIS	Marsh Pennywort	Very rare; bogs & wet places
Hyoscyamus albus *	White Henbane	Pre-1926 (Slough); rubbish tip
HYOSCYAMUS NIGER	Henbane	VU; rare; fields & waste places; sporadic
HYPERICUM ANDROSAEMUM	Tutsan	Very rare; woods in the south; also a frequent garden escape
Hypericum calycinum	Rose-of-Sharon	Uncommon; hedges & banks near habitation
HYPERICUM ELODES	Marsh St John's-wort	Very rare; boggy ponds in the south; *e.g.* Langley Park
Hypericum forrestii	Forrest's Tutsan	1997 (Hardwick); roadside
HYPERICUM HIRSUTUM	Hairy St John's-wort	Common; damp woodland rides & grassland
HYPERICUM HUMIFUSUM	Trailing St John's-wort	Rare; dry, sandy places
HYPERICUM MACULATUM SSP. OBTUSIUSCULUM	Imperforate St John's-wort	Scarce; grassy roadsides & banks
HYPERICUM MONTANUM	Pale St John's-wort	NT; very rare; woods in the south
HYPERICUM PERFORATUM	Perforate St John's-wort	Common; roadsides, disused railways & grassland
HYPERICUM PULCHRUM	Slender St John's-wort	Uncommon; woods & heathy places
HYPERICUM TETRAPTERUM	Square-stalked St John's-wort	Common; fens, wet grassland & woods
HYPERICUM X DESETANGSII	Des Etangs' St John's-wort	Very rare; roadsides & disused railways; *e.g.* Buckingham
Hypericum x inodorum	Tall Tutsan	1999 (King's Wood); woodland edge
Hypochaeris glabra *	Smooth Cat's-ear	VU; 1903 (Brickhill); sandy or gravelly heaths
HYPOCHAERIS RADICATA	Cat's-ear	Very common; grassland
Hyssopus officinalis	Hyssop	1968 (Aylesbury); waste places
IBERIS AMARA	Wild Candytuft	VU; NS; very rare; chalk grassland & field margins
Iberis umbellata	Garden Candytuft	Rare; waste places
ILEX AQUIFOLIUM	Holly	Very common; woods, hedgerows & churchyards
Ilex x altaclarensis	Highclere Holly	1998 (Windsor Hill); road verge & woodland
Impatiens capensis	Orange Balsam	Uncommon; river & canalsides, rarely by ponds
Impatiens glandulifera	Indian Balsam	Scarce; riverbanks & waste places; increasing
Impatiens parviflora	Small Balsam	Rare; open woods, woodyards & gardens
Impatiens sultanii	Busy-Lizzie	1998 (Beaconsfield); pavement cracks
Inula conyzae	Ploughman's-spikenard	Scarce; grassland & scrub on calcareous soils
INULA HELENIUM	Elecampane	1976 (Preston Bissett); ditches & orchard
Ionopsidium acaule	Violet Cress	1978 (Haddenham); garden weed
Ipomoea hederacea	Ivy-leaved Morning-glory	2004 (Lane End); maize field
IRIS FOETIDISSIMA	Stinking Iris	Rare; woods; also frequently naturalised, especially in churchyards
Iris germanica	Bearded Iris	Rare; waste places
Iris orientalis	Turkish Iris	1999 (between Mursley & Little Horwood); disused railway
IRIS PSEUDACORUS	Yellow Iris	Common; margins of rivers, canals, ditches & ponds
Iris sibirica	Siberian Iris	1992 (New Bradwell); pathside
Iris versicolor	Purple Iris	Very rare; lakes & streams; *e.g.* Caldecotte Lake
Isatis tinctoria	Woad	Pre-1926; arable fields & railwaysides; now occasionally planted
ISOLEPIS SETACEA	Bristle Club-rush	Very rare; boggy grassland on sandy soils
JASIONE MONTANA	Sheep's-bit	Very rare; sandy places; *e.g.* Bow Brickhill

Juglans nigra	Black Walnut	2004 (Old Wolverton); churchyard
Juglans regia	Walnut	Very rare; well established at Grangelands; planted elsewhere
JUNCUS ACUTIFLORUS	Sharp-flowered Rush	Uncommon; wet grassland & fens
JUNCUS AMBIGUUS	Frog Rush	1993 (Dorney Common); ditches
JUNCUS ARTICULATUS	Jointed Rush	Common; wet grassland & woodland rides
JUNCUS BUFONIUS	Toad Rush	Common; damp, bare places
JUNCUS BULBOSUS	Bulbous Rush	Very rare; boggy areas & acid pools; especially in the south
JUNCUS COMPRESSUS	Round-fruited Rush	NT; rare; meadows & disused gravel & chalk pits
JUNCUS CONGLOMERATUS	Clustered Rush	Common; woods & marshes on more acid soils
JUNCUS EFFUSUS	Soft-rush	Very common; woods, wet grassland, ponds & by water
JUNCUS INFLEXUS	Hard Rush	Very common; woods, damp grassland & by water
JUNCUS SQUARROSUS	Heath Rush	Very rare; heathy places; especially in the south
JUNCUS SUBNODULOSUS	Blunt-flowered Rush	Very rare; base-rich fens; especially in the nort
Juncus tenuis	Slender Rush	Rare; sandy woodland rides & lake margin; especially in the south
JUNCUS X DIFFUSUS		1994 (Whaddon); pond margin
Juncus x surrejanus *		1900 (Heath); marshy ground
JUNIPERUS COMMUNIS SSP. COMMUNIS	Juniper	Very rare; chalk downland
Kerria japonica	Kerria	Rare; waste places
KICKXIA ELATINE	Sharp-leaved Fluellin	Uncommon; arable fields on calcareous soils
KICKXIA SPURIA	Round-leaved Fluellin	Uncommon; arable fields on calcareous soils
KNAUTIA ARVENSIS	Field Scabious	Common; roadsides & chalk grassland
Kniphofia uvaria	Red-hot-poker	1998 (near Linford Wood); grassy bank
KOELERIA MACRANTHA	Crested Hair-grass	Scarce; old grassland on chalk or limestone
Laburnum anagyroides	Laburnum	Uncommon; planted in hedges & gardens; occasionally self-sown
Laburnum x watereri	Voss's Laburnum	Uncommon; planted in hedges & gardens; occasionally self-sown
Lactuca sativa *	Garden Lettuce	1949 (Burnham Beeches); waste places
LACTUCA SERRIOLA	Prickly Lettuce	Common; waste places; increasing
LACTUCA VIROSA	Great Lettuce	Uncommon; waste places; increasing & spreading northwards
Lagarosiphon major	Curly Waterweed	Very rare; ponds & water-filled pits
Lamiastrum galeobdolon ssp. argentatum		Common; woods, hedges & churchyards; increasing
LAMIASTRUM GALEOBDOLON SSP. MONTANUM	Yellow Archangel	Common; woods & hedges; not on clay soils
LAMIUM ALBUM	White Dead-nettle	Very common; roadsides, fields & waste places
LAMIUM AMPLEXICAULE	Henbit Dead-nettle	Scarce; cultivated ground & bases of walls; on light soils
LAMIUM HYBRIDUM	Cut-leaved Dead-nettle	Rare; disturbed ground; especially in the south
Lamium maculatum	Spotted Dead-nettle	Uncommon; woods, hedges & waste places
LAMIUM PURPUREUM	Red Dead-nettle	Very common; cultivated ground
Lappula squarrosa *	Bur Forget-me-not	1917 (Naphill); waste places; introduced with seed
LAPSANA COMMUNIS SSP. COMMUNIS	Nipplewort	Very common; waste places & open woods
Larix decidua	European Larch	Frequent; woods & hedges
Larix kaempferi	Japanese Larch	Rare; woods; probably under-recorded
Larix x marschlinsii	Hybrid Larch	Rare; woods; probably under-recorded
Lathraea clandestina	Purple Toothwort	1989 (Medmenham); on poplar roots by rive
LATHRAEA SQUAMARIA	Toothwort	Very rare; hedgerows & open woodland; *e.g.* Wendover Park
Lathyrus aphaca	Yellow Vetchling	VU; very rare; disturbed ground & open grassland; *e.g.* Stonepit Field
Lathyrus cicera *	Red Vetchling	1957 (Farnham); waste ground
Lathyrus grandiflorus	Two-flowered Everlasting-pea	Very rare; churchyards; *e.g.* Turville
Lathyrus hirsutus *	Hairy Vetchling	Pre-1926 (Old Linslade & Iver Heath); waste ground & clover field
Lathyrus latifolius	Broad-leaved Everlasting-pea	Scarce; hedges & waste places
LATHYRUS LINIFOLIUS	Bitter-vetch	Very rare; woods on acid soils in the south
LATHYRUS NISSOLIA	Grass Vetchling	Scarce; grassy places; maybe overlooked
Lathyrus odoratus	Sweet Pea	1997 (Heelands & Conniburrow); waste places
LATHYRUS PRATENSIS	Meadow Vetchling	Common; roadsides, open woodland rides & grassy places
LATHYRUS SYLVESTRIS	Narrow-leaved Everlasting-pea	Very rare; woodland edges & hedges in the north
Lathyrus tuberosus	Tuberous Pea	1987 (Sands Bank); calcareous grassland
Laurus nobilis	Bay	1986 (Chalfont Grove); woodland
Lavandula angustifolia *s.l.*	Lavender	Very rare; frequently planted; rarely self-sown

Lavatera x clementii	Garden Tree-mallow	1999 (Little Marlow); roadsides & waste ground
Lavatera trimestris	Royal Mallow	2003 (Lane End); waste places
LEGOUSIA HYBRIDA	Venus's-looking-glass	Scarce; arable fields on calcareous soils; decreasing
LEMNA GIBBA	Fat Duckweed	Scarce; rivers & ponds
LEMNA MINOR	Common Duckweed	Common; rivers, canals, ditches & ponds
Lemna minuta	Least Duckweed	Uncommon; ponds; recently introduced, rapidly increasing
LEMNA TRISULCA	Ivy-leaved Duckweed	Uncommon; ponds
Lens culinaris *	Lentil	Pre-1926 (Iver); rubbish heap
LEONTODON AUTUMNALIS SSP. AUTUMNALIS	Autumn Hawkbit	Common; grassland; mostly on calcareous soils
LEONTODON HISPIDUS	Rough Hawkbit	Common; grassland; mostly on calcareous soils
LEONTODON SAXATILIS	Lesser Hawkbit	Uncommon; grassland on light soils
Leonurus cardiaca	Motherwort	Very rare; garden weed at Stoke Poges
LEPIDIUM CAMPESTRE	Field Pepperwort	Scarce; rough or bare ground
Lepidium draba ssp. chalepense	Lens-podded Hoary Cress	Very rare; roadside at Old Slade
Lepidium draba ssp. draba	Hoary Cress	Scarce; roadsides; possibly decreasing
LEPIDIUM HETEROPHYLLUM	Smith's Pepperwort	Very rare; sandy places; *e.g.* Langley Park
Lepidium latifolium	Dittander	Very rare; rough grassland in the south
Lepidium perfoliatum *	Perfoliate Pepperwort	Pre-1926 (Iver); waste places
LEPIDIUM RUDERALE	Narrow-leaved Pepperwort	Very rare; roadsides & waste places
Lepidium sativum	Garden Cress	1995 (Emberton); waste ground & garden
Lepidium virginicum *	Least Pepperwort	1928 (Beaconsfield); waste places
LEUCANTHEMUM VULGARE	Oxeye Daisy	Very common; roadsides & grassland
Leucanthemum x superbum	Shasta Daisy	Rare; waste places; usually close to habitation
LEUCOJUM AESTIVUM SSP. AESTIVUM	Summer Snowflake	RDB; very rare; meadows & carr by the River Thames; planted elsewhere
Leucojum vernum	Spring Snowflake	1986 (Hollywell Plantation); woodland
Leycesteria formosa	Himalayan Honeysuckle	Very rare; churchyards, gardens & scrub
Libertia formosa	Chilean-iris	Very rare; churchyard at Drayton Parslow
Ligularia dentata	Leopardplant	Very rare; parkland; *e.g.* Green Park
Ligustrum ovalifolium	Garden Privet	Frequent; hedges; probably always planted
LIGUSTRUM VULGARE	Wild Privet	Common; hedges & scrub on calcareous soils
Lilium martagon	Martagon Lily	Very rare; parkland; *e.g.* Stowe Park
Lilium x hollandicum	Hybrid Orange-lily	1989 (Loughton); road verge
Limnanthes douglasii	Meadow-foam	1994 (Tongwell Lake); lake margin
LIMOSELLA AQUATICA	Mudwort	NS; Very rare; muddy edges of reservoirs & ponds; *e.g.* Startop's End
Linaria maroccana	Annual Toadflax	1996 (Weston Underwood); waste places
Linaria purpurea	Purple Toadflax	Common; walls & waste places near habitation
LINARIA REPENS	Pale Toadflax	Rare; grassland on calcareous soils; *e.g.* Bradenham churchyard
LINARIA VULGARIS	Common Toadflax	Common; roadsides, walls & railway banks
Linaria x dominii		1998 (near Wendover); roadside
LINARIA X SEPIUM		1998 (Stoke Hammond); roadside
Linum bienne *	Pale Flax	1954 (Fingest); roadside
LINUM CATHARTICUM	Fairy Flax	Locally common; short grass; mostly on calcareous soils
Linum usitatissimum	Flax	Rare; roadsides & disturbed ground; often of birdseed origin
Liquidambar styraciflua	Sweet Gum	Very rare; planted in Hockeridge & Pancake Woods
Liriodendron tulipifera	Tulip Tree	Very rare; planted in Hockeridge & Pancake Woods
LISTERA OVATA	Common Twayblade	Uncommon; woodland & open calcareous soil
LITHOSPERMUM ARVENSE	Field Gromwell	EN; very rare; arable fields & waste places
LITHOSPERMUM OFFICINALE	Common Gromwell	Very rare; woodland & scrub; *e.g.* Salcey Forest & Hambledon
LITTORELLA UNIFLORA	Shoreweed	Very rare; pond margin & gravel pit; *e.g.* Marlow
Lobelia erinus	Garden Lobelia	Uncommon; walls & pavement cracks
Lobularia maritima	Sweet Alison	Uncommon; walls & pavement cracks
Lolium multiflorum	Italian Rye-grass	Uncommon; arable fields & grass leys
LOLIUM PERENNE	Perennial Rye-grass	Very common; grasslands; much planted
Lolium persicum *	Persian Darnel	1968 (Aylesbury); rubbish heap
Lolium remotum *	Flaxfield Rye-grass	1906 (Cold Brayfield); waste ground
Lolium rigidum	Mediterranean Rye-grass	1973 (Ashley Green); roadside
Lolium temulentum *	Darnel	CR; 1967 (New Bradwell); waste ground
Lolium x boucheanum	Hybrid Rye-grass	Very rare; arable fields & waste places; *e.g.* Wing
Lonicera caprifolium *	Perfoliate Honeysuckle	Pre-1926 (Harleyford); woods & hedges
Lonicera henryi	Henry's Honeysuckle	1997 (Heelands); scrub
Lonicera japonica	Japanese Honeysuckle	Rare; hedges & railway banks; *e.g.* Central Milton Keynes

Aaron Woods

Ludwigia grandiflora
(Uruguayan Hampshire-purslane)
[Chesham Bois Common]

Aaron Woods

Oxalis corniculata var. atropurpurea (Procumbent Yellow-sorrel)
[Bletchley]

Diana Stroud

Hyoscyamus niger (Henbane) [Dorney Common]

Roy Maycock

Downingia elegans (Californian Lobelia)
[Willen]

Roy Maycock

Nicotiana rustica (Wild Tobacco) [Great Missenden]

Drosera intermedia (Oblong-leaved Sundew) [Stoke Common]

Ophrys insectifera (Fly Orchid) [Homefield Wood]

Damasonium alisma (Starfruit) [Naphill Common]

Ophrys apifera var. trollii (Wasp Orchid) [Broughton]

Cyperus fuscus (Brown Galingale) [Dorney Common]

Gentianella anglica (Early Gentian)
[Ivinghoe Beacon]

Roy Maycock

Dapnhe mezereum (Mezereon)
[Aston Clinton]

Roy Maycock

Lamium amplexicaule (Henbit Dead-nettle) [Bletchley]

Roy Maycock

Scandix pecten-veneris (Shepherd's-needle)
[Whelpley Hill]

Roy Maycock

Myosurus minimus (Mousetail)
[Long Herdon Meadow]

Julia Carey

Veronica chamaedrys (Germander Speedwell)
[Grangelands]

Roy Maycock

Dipsacus pilosus (Small Teasel)
[Bozenham Mill]

Roy Maycock

Alchemilla xanthochlora (Intermediate Lady's-mantle)
[Halton Wood]

Roy Maycock

Cichorium intybus (Chicory) [Marlow Bottom]

Roy Maycock

Vulpia ciliata ssp. ambigua (Bearded Fescue),
Sedium acre (Biting Stonecrop)
& *Catapodium rigidumo* (Fern-grass)
[Olney]

Lonicera nitida	Wilson's Honeysuckle	Frequent; hedges; probably always planted
LONICERA PERICLYMENUM	Honeysuckle	Common; woods & hedges
Lonicera pileata	Box-leaved Honeysuckle	Rare; much planted in urban areas; frequently self-sown
Lonicera tatarica	Tatarian Honeysuckle	Very rare; roadside at Hughenden Valley
Lonicera xylosteum *	Fly Honeysuckle	RDB; 1930s (Aylesbury); ornamental woods & parkland
LOTUS CORNICULATUS	Common Bird's-foot-trefoil	Common; grasslands, roadsides & churchyards
LOTUS GLABER	Narrow-leaved Bird's-foot-trefoil	Rare; bare clay & sparse grassland
LOTUS PEDUNCULATUS	Greater Bird's-foot-trefoil	Common; wet grassland & wet woodland rides
Ludwigia grandiflora	Uruguayan Hampshire-purslane	Very rare; pond at Chesham Bois Common
Lunaria annua	Honesty	Common; roadsides & hedgebanks; especially near habitation
Lupinus arboreus	Tree Lupin	Very rare; sandy heathland at New Wavendon Heath
Lupinus x regalis	Russell Lupin	Very rare; waste places; *e.g.* Iver
LUZULA CAMPESTRIS	Field Wood-rush	Common; lawns, churchyards & heaths
LUZULA FORSTERI	Forster's Wood-rush	Scarce; woods; especially on The Chilterns
Luzula luzuloides	White Wood-rush	1986 (Cliveden); woodland
LUZULA MULTIFLORA SSP. CONGESTA		Scarce; woods & grassland on acid soils
LUZULA MULTIFLORA SSP. MULTIFLORA	Heath Wood-rush	Rare; woods & grassland on acid soils
LUZULA PILOSA	Hairy Wood-rush	Locally common; woods; especially on The Chilterns
LUZULA SYLVATICA	Great Wood-rush	Rare; heathy woods; *e.g.* Whitfield Wood
LUZULA X BORRERI	Hybrid Wood-rush	Very rare; woods on The Chilterns; *e.g.* Kings Wood
Lychnis coronaria	Rose Campion	Rare; waste places
LYCHNIS FLOS-CUCULI	Ragged-Robin	Uncommon; fens, wet grassland & wet woodland rides
Lycium barbarum s.l.	Duke of Argyll's Teaplant	Rare; hedges & walls near habitation
Lycopersicon esculentum	Tomato	Rare; rubbish dumps, sewage waste & waste places
Lycopodiella inundata *	Marsh Clubmoss	EN; 1946 (New Wavendon Heath); bogs & heaths
Lycopodium clavatum	Stag's-horn Clubmoss	1974 (Hillock Wood); heathy woods
LYCOPUS EUROPAEUS	Gypsywort	Common; rivers, canals & ponds
Lysichiton americanus	American Skunk-cabbage	Very rare; bogs & streams in the south; *e.g.* Burnham Beeches
Lysimachia ciliata	Fringed Loosestrife	1970 (Iver Heath); rough ground
LYSIMACHIA NEMORUM	Yellow Pimpernel	Uncommon; woods; not on alkaline soils
LYSIMACHIA NUMMULARIA	Creeping-Jenny	Common; marshy grassland, churchyards & woods
Lysimachia punctata	Dotted Loosestrife	Uncommon; roadsides
Lysimachia thyrsiflora *	Tufted Loosestrife	1931 (Stoke Poges); ponds
LYSIMACHIA VULGARIS	Yellow Loosestrife	Uncommon; riverbanks & marshes
Lythrum hyssopifolium	Grass-poly	RDB; very rare; lake margins at Willen
Lythrum junceum	False Grass-poly	1970 (Aylesbury); rubbish dump
LYTHRUM PORTULA	Water-purslane	Very rare; muddy edges of acid ponds; *e.g.* Naphill Common
LYTHRUM SALICARIA	Purple-loosestrife	Locally common; rivers, canals, ponds & wet woods
Macleaya x kewensis	Plume-poppy	1992 (Mentmore); churchyard
Mahonia aquifolium	Oregon-grape	Uncommon; woods, hedges & churchyards
Malcolmia africana *	African Stock	Pre-1926 (Slough); waste ground
Malcolmia maritima	Virginia Stock	1994 (Willen Priory); waste ground
Malus floribunda	Japanese Crab	1996 (near Hambledon); roadside
Malus 'John Downie'		1996 (Medmenham); hedge
Malus pumila	Apple	Common; hedges
MALUS SYLVESTRIS	Crab Apple	Rare; woods & hedges; much over-recorded
Malus x purpurea	Purple Crab	1999 (Drayton Beauchamp); churchyard
Malva alcea	Greater Musk-mallow	1998 (North Crawley); churchyard
MALVA MOSCHATA	Musk-mallow	Uncommon; grassy & waste places
MALVA NEGLECTA	Dwarf Mallow	Uncommon; roadsides, churchyards & waste places
Malva parviflora *	Least Mallow	Pre-1926 (Slough & Linslade); waste places
Malva pusilla *	Small Mallow	*c.*1941 (near High Wycombe); waste places
MALVA SYLVESTRIS	Common Mallow	Common; waste places & roadsides
Marrubium vulgare *	White Horehound	NS; pre-1926; roadsides & waste places; especially in the south
Matricaria discoidea	Pineappleweed	Very common; trodden places *e.g.* paths & gateways
MATRICARIA RECUTITA	Scented Mayweed	Common; arable fields, roadsides & waste places
Meconopsis cambrica	Welsh Poppy	NS; bases of walls in villages
Medicago arabica	Spotted Medick	Scarce; short turf; increasing
MEDICAGO LUPULINA	Black Medick	Very common; short turf & waste places
Medicago minima *	Bur Medick	Pre-1961 (Unlocalised); wool alien

Medicago polymorpha	Toothed Medick	NS; very rare; waste places; *e.g.* near Marlow
Medicago sativa ssp. falcata	Sickle Medick	1975 (Iver); waste ground
Medicago sativa ssp. sativa	Lucerne	Uncommon; roadsides & field margins
Medicago sativa ssp. varia *	Sand Lucerne	Pre-1985 (Micklefield); waste ground
Melampyrum cristatum *	Crested Cow-wheat	NS; 1769 (Moreton Green); shaded bank
MELAMPYRUM PRATENSE SSP. PRATENSE	Common Cow-wheat	Rare; woods on acid soils in the south; decreasing
MELICA UNIFLORA	Wood Melick	Locally common; woods & hedgerows on light soils
Melilotus albus	White Melilot	Rare; roadsides & waste places; especially in the south
MELILOTUS ALTISSIMUS	Tall Melilot	Uncommon; rough grassy places
Melilotus indicus	Small-flowered Melilot	Very rare; waste places; *e.g.* Lane End
Melilotus officinalis	Ribbed Melilot	Uncommon; rough grassy places
Melilotus sulcatus *	Furrowed Melilot	1916 (Slough); waste ground
Melissa officinalis	Balm	Rare; walls & grassy places
MENTHA AQUATICA	Water Mint	Common; wet places & margins of waterbodies
MENTHA ARVENSIS	Corn Mint	Uncommon; damp meadows & open woodland rides
MENTHA PULEGIUM	Pennyroyal	EN; RDB; pre-1926; extinct as a native; now a rare garden escape
MENTHA SPICATA	Spear Mint	Uncommon; waste places near habitation
Mentha suaveolens	Round-leaved Mint	Very rare; waste places near habitation
Mentha x gracilis *	Bushy Mint	1897 (near West Wycombe); waste places
Mentha x piperita	Peppermint	Rare; wet places, ditches & ponds
Mentha x rotundifolia *	False Apple-mint	1964 (Bellingdon); waste ground
Mentha x smithiana	Tall Mint	1974 (High Wycombe); wet places
MENTHA X VERTICILLATA	Whorled Mint	Rare; wet places; not always with parents
Mentha x villosa	Apple-mint	Scarce; roadsides & waste places; *e.g.* Ravenstone
Mentha x villosonervata	Sharp-toothed Mint	1997 (Great Missenden); hedge
MENYANTHES TRIFOLIATA	Bogbean	Very rare; ponds on acid substrates in the south; *e.g.* Burnham Beeches
MERCURIALIS ANNUA	Annual Mercury	Scarce; cultivated ground on light soils; increasing
MERCURIALIS PERENNIS	Dog's Mercury	Common; broadleaved woodlands; especially on heavy soils
MESPILUS GERMANICA	Medlar	Very rare; hedges & river bank; *e.g.* Stony Stratford
Metasequoia glyptostroboides	Dawn Redwood	1989 (Pancake Wood); woodland
MILIUM EFFUSUM	Wood Millet	Locally common; woods; especially on calcareous soils
Mimulus guttatus	Monkeyflower	Rare; streamsides; *e.g.* Chesham & High Wycombe
Mimulus moschatus	Musk	1974 (Chenies); riverbank
Mimulus x burnetii	Coppery Monkeyflower	1997 (Princes Risborough); waste ground
Mimulus x robertsii	Hybrid Monkeyflower	1991 (Aylesbury); pond margin
MINUARTIA HYBRIDA SSP. TENUIFOLIA	Fine-leaved Sandwort	EN; NS; 1985 (Bradenham); old walls & chalk
Mirabilis jalapa	Marvel-of-Peru	2001 (Stoke Mandeville); dung heap
MISOPATES ORONTIUM	Weasel's-snout	VU; very rare; disturbed ground on light soils; *e.g.* Iver Heath
MOEHRINGIA TRINERVIA	Three-nerved Sandwort	Common; woods & shady places
MOENCHIA ERECTA	Upright Chickweed	1988 (Church Wood); sandy turf
MOLINIA CAERULEA SSP. ARUNDINACEA		Very rare; fens; *e.g.* Pilch Fields
MOLINIA CAERULEA SSP. CAERULEA	Purple Moor-grass	Rare; fens & bogs
Molucella laevis	Bells-of-Ireland	1970 (Langley Park); rubbish heap
Monotropa hypopitys	Yellow Bird's-nest	EN; rare; woods; both sspp. *hypophega* & *hypopitys* occur but distributions not known
Montia fontana	Blinks	Rare; wet meadows & woods; sspp. *amphoritana*, *chondrosperma* & *variabilis* are all said to occur but distributions not known
Morina longifolia	Whorlflower	Very rare; allotment at Soulbury
Morus nigra	Mulberry	1995 (Medmenham); plante
Muscari armeniacum	Garden Grape-hyacinth	Uncommon; churchyards, roadsides & hedgebanks; increasing
Muscari azureum	Azure Hyacinth	Very rare; churchyard at Halton
Muscari botryoides	Compact Grape-hyacinth	Very rare; churchyard at Ellesborough
Muscari neglectum	Grape-hyacinth	RDB; rare; churchyards & roadsides
MYCELIS MURALIS	Wall Lettuce	Uncommon; woods & walls on calcareous soils in the south
MYOSOTIS ARVENSIS	Field Forget-me-not	Common; cultivated land & woods
MYOSOTIS DISCOLOR	Changing Forget-me-not	Scarce; dry sandy soils & marshes; widespread
Myosotis dissitiflora *	Perennial Forget-me-not	1914 (Taplow); waste ground
MYOSOTIS LAXA SSP. CAESPITOSA	Tufted Forget-me-not	Uncommon; sides of ponds & marshy places

Species	Common Name	Status & Habitat
MYOSOTIS RAMOSISSIMA	Early Forget-me-not	Rare; dry sandy soils & arable fields
MYOSOTIS SCORPIOIDES	Water Forget-me-not	Common; margins of waterbodies & wet grassland
MYOSOTIS SECUNDA	Creeping Forget-me-not	Very rare; wet heathy places & pond margins
Myosotis sylvatica	Wood Forget-me-not	Common; churchyards, hedges & woods
MYOSOTON AQUATICUM	Water Chickweed	Uncommon; by ponds & rivers; occasionally in wet grassland
MYOSURUS MINIMUS	Mousetail	VU; very rare; damp barish fields
Myrica gale *	Bog Myrtle	1597 (Colnbrook); damp heathy places
MYRIOPHYLLUM ALTERNIFLORUM	Alternate Water-milfoil	Very rare; ponds on acid soils in the south; e.g. Naphill Common
Myriophyllum aquaticum	Parrot's-feather	Very rare; ponds & lakes
MYRIOPHYLLUM SPICATUM	Spiked Water-milfoil	Rare; ponds, lakes & canals with clean water
MYRIOPHYLLUM VERTICILLATUM	Whorled Water-milfoil	NS; very rare; lakes, rivers & ponds; e.g. Stowe Park
Myrrhis odorata	Sweet Cicely	Very rare; waste places; e.g. near Stockgrove
Narcissus bicolor	Two-coloured Daffodil	1997 (New Bradwell); disused railwaybank
Narcissus cyclamineus	Cyclamen-flowered Daffodil	1994 (Thornborough); green lane
Narcissus nobilis	Large-flowered Daffodil	1997 (Addington); roadside
Narcissus poeticus ssp. poeticus	Pheasant's-eye Daffodil	Rare; roadsides & grassy places; e.g. Tathall End
Narcissus pseudonarcissus ssp. major	Spanish Daffodil	1997 (Bletchley); rough grassland
Narcissus pseudonarcissus ssp. obvallaris	Tenby Daffodil	Very rare; parkland at Campbell Park
NARCISSUS PSEUDONARCISSUS SSP. PSEUDONARCISSUS	Daffodil	Very rare; woods & churchyards; e.g. Bow Brickhill
Narcissus x boutigyanus	White-and-orange Daffodil	Very rare; churchyards & grassy places; e.g. Stantonbury
Narcissus x incomparabilis	Nonesuch Daffodil	Very rare; churchyards & grassy places; e.g. Joan's Piece
Narcissus x medioluteus	Primrose-peerless	Very rare; grassy places; e.g. Linford Wood
NARDUS STRICTA	Mat-grass	Very rare; poor acid grassland & heaths; e.g. Moorend Common
Narthecium ossifragum *	Bog Asphodel	1959 (East Burnham Common); bogs on acid soils
Nemesia strumosa *	Cape-jewels	1912 (Pitstone Hill); calcareous grassland
NEOTTIA NIDUS-AVIS	Bird's-nest Orchid	NT; very rare; beech & hazel woods; decreasing
NEPETA CATARIA	Cat-mint	VU; very rare; roadsides & hedgerows; much decreased
Nepeta x faassenii	Garden Cat-mint	Rare; waste places near habitation; e.g. Ravenstone
Neslia paniculata *	Ball Mustard	Pre-1926 (Slough & Taplow); waste places
Nicandra physalodes	Apple-of-Peru	Very rare; disturbed ground; e.g. Aston Abbotts
Nicotiana rustica	Wild Tobacco	1997 (Great Missenden); rubbish heap
Nicotiana x sanderae	Flowering Tobacco	1996 (Dibden Hill); compost heap
Nigella damascena	Love-in-a-mist	Very rare; waste places near habitation
Nonea lutea	Yellow Nonea	Very rare; garden weed at Bletchley
Nonea rosea *	Pink Nonea	1911 (Low Scrubs); site of garden
Nothofagus antarctica	Antarctic Beech	Pre-1985 (Philipshill Wood); woodland
Nothofagus fusca	Red Beech	Very rare; planted in Park Wood
Nothofagus nervosa	Rauli	Very rare; planted in woods; e.g. Linford Wood
Nothofagus obliqua	Roble	Very rare; planted in a few woods
NUPHAR LUTEA	Yellow Water-lily	Scarce; rivers, canals, ponds & lakes
NYMPHAEA ALBA SSP. ALBA	White Water-lily	Rare; rivers & ponds; decreasing; often introduce
NYMPHOIDES PELTATA	Fringed Water-lily	NS; rare; rivers, ponds & gravel pits; often introduced
ODONTITES VERNUS SSP. SEROTINUS	Red Bartsia	Common; tracks & grassy places
OENANTHE AQUATICA	Fine-leaved Water-dropwort	Very rare; ponds; e.g. Fulmer & Grendon Underwood
OENANTHE CROCATA	Hemlock Water-dropwort	Rare; rivers, canals & ditches; mostly in the south
OENANTHE FISTULOSA	Tubular Water-dropwort	VU; rare; wet grassland; especially in the Ray & Thame valleys
OENANTHE FLUVIATILIS	River Water-dropwort	Very rare; rivers; e.g. Stony Stratford; much decreased
OENANTHE PIMPINELLOIDES	Corky-fruited Water-dropwort	Very rare; meadows & old orchard; now only known from Prestwood
OENANTHE SILAIFOLIA	Narrow-leaved Water-dropwort	NT; NS; very rare; water meadows in the Ray valley
Oenothera biennis	Common Evening-primrose	Very rare; waste places & railway ballast; e.g. Bletchley

Oenothera glazioviana	Large-flowered Evening-primrose	Uncommon; waste places; increasing
Oenothera x fallax	Intermediate Evening-primrose	1991 (Aylesbury); site of old allotments
Omphalodes verna *	Blue-eyed-Mary	Pre-1926 (Weston Underwood & Brill); copses & plantations
Onobrychis viciifolia	Sainfoin	NT; scarce; grassy places on calcareous soils; especially in The Chilterns
Onoclea sensibilis	Sensitive Fern	1985 (Cliveden); parkland
ONONIS REPENS SSP. REPENS	Common Restharrow	Uncommon; dry grassland; especially on calcareous soils
ONONIS SPINOSA	Spiny Restharrow	Scarce; fields & roadsides; commoner in the north
Onopordum acanthium	Scotch Thistle	Uncommon; roadsides & waste places; *e.g.* near Lathbury
OPHIOGLOSSUM VULGATUM	Adder's-tongue	Rare; old grassland; decreasing
OPHRYS APIFERA	Bee Orchid	Uncommon; grassland & bare clay; often where soil has been disturbed; increasing
OPHRYS INSECTIFERA	Fly Orchid	VU; very rare; woods & grassy hedgebanks on the chalk
ORCHIS MASCULA	Early-purple Orchid	Scarce; woodland; commoner in the north
ORCHIS MILITARIS	Military Orchid	VU; RDB; very rare; woodland & scrub; now only in two south Chilterns sites
ORCHIS MORIO	Green-winged Orchid	NT; rare; old meadows; much decreased
Orchis ustulata *	Burnt Orchid	1961 (Coombe Hill); chalk grassland
OREOPTERIS LIMBOSPERMA	Lemon-scented Fern	2002 (Black Park); heathy woods
Origanum virens	Green Oregano	1973 (Sawyers Green); waste ground
ORIGANUM VULGARE	Marjoram	Locally common; calcareous grassland
Ornithogalum angustifolium	Star-of-Bethlehem	Scarce; grassy places, roadsides & railway banks
Ornithogalum nutans	Drooping Star-of-Bethlehem	Very rare; riverbank & churchyard; *e.g.* Broughton
Ornithogalum pyrenaicum	Spiked Star-of-Bethlehem	NS; 1998 (Snelshall East); bank of Loughton Brook
ORNITHOPUS PERPUSILLUS	Bird's-foot	Rare; dry sandy heaths
OROBANCHE ELATIOR	Knapweed Broomrape	Very rare; roadside & grassy places
OROBANCHE MINOR	Common Broomrape	Rare; grassy places, clover fields & shrub beds; increasing on planted *Brachyglottis* 'Sunshine'
Orobanche picridis *	Oxtongue Broomrape	EN; 1955 (Ivinghoe); chalk grassland
Orobanche rapum-genistae *	Greater Broomrape	NT; NS; 1950 (Amerden); heaths & commons
OSMUNDA REGALIS	Royal Fern	Very rare; bogs in peaty woods; *e.g.* Old Wavendon Heath
OXALIS ACETOSELLA	Wood-sorrel	Locally common; woods & hedgebanks
Oxalis articulata ssp. articulata	Pink-sorrel	Scarce; grassy & waste places
Oxalis corniculata	Procumbent Yellow-sorrel	Frequent; pavement cracks, walls & gardens
Oxalis debilis	Large-flowered Pink-sorrel	Very rare; churchyard & allotments; *e.g.* Stoke Goldington
Oxalis exilis	Least Yellow-sorrel	Rare; cultivated ground, pavement cracks & lawns
Oxalis incarnata	Pale Pink-sorrel	Rare; garden weed; *e.g.* Bradwell Village
Oxalis latifolia ssp. latifolia	Garden Pink-sorrel	1991 (Aylesbury); disturbed ground
Oxalis stricta	Upright Yellow-sorrel	Rare; garden weed; *e.g.* Bletchley
Pachysandra terminalis	Carpet Box	1999 (Central Milton Keynes); shrubbery
Paeonia officinalis	Garden Peony	Very rare; wood & abandoned garden; *e.g.* Great Tinkers Wood
Panicum capillare	Witch-grass	1970 (Iver); rubbish heap
Panicum miliaceum	Common Millet	Rare; waste places; usually of birdseed origin
PAPAVER ARGEMONE	Prickly Poppy	VU; very rare; arable fields on light or chalky soils
Papaver atlanticum	Atlas Poppy	Very rare; walls & waste places near habitation
PAPAVER DUBIUM SSP. DUBIUM	Long-headed Poppy	Common; arable fields, waste places & walls
PAPAVER DUBIUM SSP. LECOQII	Yellow-juiced Poppy	Scarce; arable fields & waste places; commoner in the north
Papaver hybridum *	Rough Poppy	1929 (Burnham Beeches); arable fields & waste places
Papaver pseudoorientale	Oriental Poppy	Very rare; waste places near habitation
PAPAVER RHOEAS	Common Poppy	Common; arable fields & waste places
Papaver somniferum ssp. setigerum *	Small-flowered Opium Poppy	Pre-1926 (Maids Moreton); waste places
Papaver somniferum ssp. somniferum	Opium Poppy	Frequent; waste places near to habitation
PAPAVER X HUNGARICUM *		1912 (Kimble); rectory garden
Parapholis pycnantha *		1903 (Iver); rubbish tip

Parentucellia viscosa	Yellow Bartsia	Very rare; grassy places; introduced with seed mix; *e.g.* Willen
PARIETARIA JUDAICA	Pellitory-of-the-wall	Scarce; old walls; often around churches
PARIS QUADRIFOLIA	Herb-Paris	Rare; dense woods; especially on calcareous soils
Parnassia palustris *	Grass-of-Parnassus	Pre-1926; marshy places; *e.g.* Winslow & Aylesbury
Parthenocissus inserta	False Virginia-creeper	1997 (Heelands); hedge
Parthenocissus quinquefolia	Virginia-creeper	Scarce; walls & hedges near habitation; *e.g.* High Wycombe
Parthenocissus tricuspidata	Boston-ivy	1982 (Haddenham); churchyard
PASTINACA SATIVA	Wild Parsnip	Uncommon; roadsides & old grassland on calcareous soils
Pedicularis palustris *	Marsh Lousewort	1963 (Fawley); wet meadows
PEDICULARIS SYLVATICA SSP. SYLVATICA	Lousewort	Very rare; wet acid soils; *e.g.* Little Brickhill
Pentaglottis sempervirens	Green Alkanet	Common; hedges, roadsides & bases of walls near habitation
PERSICARIA AMPHIBIA	Amphibious Bistort	Common; rivers, canals, ponds & their margins
Persicaria amplexicaulis	Red Bistort	Very rare; hedge, pond & churchyard; *e.g.* Prestwood churchyard
PERSICARIA BISTORTA	Common Bistort	Very rare; damp grassland; sometimes planted
PERSICARIA HYDROPIPER	Water-pepper	Common; pond margins & damp ruts in woodland rides
PERSICARIA LAPATHIFOLIA	Pale Persicaria	Common; arable fields & waste places
PERSICARIA MACULOSA	Redshank	Common; arable fields & damp places
PERSICARIA MINOR	Small Water-pepper	VU; very rare; damp places on commons; *e.g.* Littleworth Common
PERSICARIA MITIS	Tasteless Water-pepper	VU; 1975 (Langley Park); sides of ditches & streams
Persicaria wallichii *	Himalayan Knotweed	1963 (High Wycombe); waste ground
Persicaria x intercedens *		Pre-1926 (Colnbrook); damp places
Petasites albus	White Butterbur	Very rare; wood & bank; *e.g.* Ballinger Dell Wood
Petasites fragrans	Winter Heliotrope	Uncommon; roadsides & hedgebanks; *e.g.* Lower End, Wavendon
PETASITES HYBRIDUS	Butterbur	Rare; by rivers, canals & ponds; female plants very rare
Petasites japonicus	Giant Butterbur	1995 (Bakers Wood)
Petrorhagia nanteuilii	Childing Pink	VU; 1978 (Dorton); railway cutting; possibly in error for *P. prolifera*
Petroselinum crispum	Garden Parsley	Rare; waste places & allotments
PETROSELINUM SEGETUM	Corn Parsley	Very rare; dry banks & cornfields; *e.g.* Calverton
Petunia x hybrida	Petunia	1995 (Woburn Sands); between paving
Peucedanum palustre *	Milk-parsley	VU; 1799 (Bulstrode Park); pond
Phacelia tanacetifolia	Phacelia	Rare; planted as a holding crop & for bees; rarely escapes
Phalaris angusta *	Timothy Canary-grass	1909 (Slough); waste places
Phalaris aquatica	Bulbous Canary-grass	Very rare; fields & waste places; *e.g.* Linslade
PHALARIS ARUNDINACEA	Reed Canary-grass	Common; watersides & wet places; occasionally planted
Phalaris canariensis	Canary-grass	Rare; roadsides & waste places; often from birdseed
Phalaris coerulescens *	Blue Canary-grass	1903 (Iver); waste places
Phalaris minor *	Lesser Canary-grass	Pre-1926 (Iver); waste places
Phaseolus coccineus	Runner Bean	1995 (Chenies); waste places
Phaseolus vulgaris *	French Bean	Pre-1926 (Langley); brickyard
Philadelphus coronarius	Mock-orange	Rare; hedges near habitation
Philadelphus x virginalis	Hairy Mock-orange	Very rare; hedges near habitation; possibly under-recorded
Phillyrea latifolia	Mock Olive	Very rare; parkland; *e.g.* Green Park & Willen Well
PHLEUM BERTOLONII	Smaller Cat's-tail	Common; grassy places
Phleum paniculatum *	British Timothy	*c.*1800 (Fulmer); field
Phleum phleoides *	Purple-stem Cat's-tail	Pre-1926 (Langley); waste places
PHLEUM PRATENSE	Timothy	Very common; grassland
Phlomis russelliana *	Turkish Sage	1968 (Kimble Rifle Range); calcareous grassland
Phoenix dactylifera	Date	1970 (Weedon); roadside
PHRAGMITES AUSTRALIS	Common Reed	Common; by water; often in large stands
Phuopsis stylosa	Caucasian Crosswort	Very rare; waste ground by Fennell's Wood
Phygelius capensis	Cape Figwort	1999 (Beaconsfield); churchyard
PHYLLITIS SCOLOPENDRIUM	Hart's-tongue	Uncommon; walls, churchyards, woods & hedgebanks; increasing
Physalis alkekengi	Japanese-lantern	Very rare; roadsides & waste places; *e.g.* Bletchley
Physalis ixocarpa	Tomatillo	1992 (New Bradwell); garden weed
Physalis peruviana	Cape-gooseberry	1997 (Colnbrook); roadside
Physalis philadelphica	Large-flowered Tomatillo	1992 (Newport Pagnell); garden weed
Physospermum cornubiense	Bladderseed	Very rare; woodland; only known from Dorney Wood
Picea abies	Norway Spruce	Common; plantations; often self-sown
Picea breweriana	Brewer Spruce	1987 (Splash Covert); woodland
Picea omorika	Serbian Spruce	1990 (Lamport Wood); plantations
Picea sitchensis	Sitka Spruce	Rare; plantations; sometimes self-sown
PICRIS ECHIOIDES	Bristly Oxtongue	Locally abundant; disturbed ground & waste places

Scientific Name	Common Name	Notes
PICRIS HIERACIOIDES SSP. HIERACIOIDES	Hawkweed Oxtongue	Uncommon; calcareous grassland & hedgerows
Pieris japonica	Japanese Fetter-bush	1984 (Penn); churchyard
Pilea microphylla	Artillery Plant	2005 (Woburn Sands); greenhouse weed
Pilosella aurantiaca ssp. carpathicola	Fox-and-cubs	Rare; lawns & waste places
Pilosella caespitosa ssp. colliniformis	Yellow Fox-and-cubs	1999 (Downley); chalk bank
Pilosella flagellaris ssp. flagellaris	Whip Mouse-ear-hawkweed	1979 (Castlethorpe); railway bank
PILOSELLA OFFICINARUM	Mouse-ear-hawkweed	Common; short calcareous turf in full sun
Pilosella praealta ssp. praealta *	Tall Mouse-ear-hawkweed	1912 (near Hanslope); railway bank
Pilosella praealta ssp. thaumasia		1978 (Twyford); disused railway bank
PIMPINELLA MAJOR	Greater Burnet-saxifrage	Uncommon; roadsides & hedgerows; especially on alkaline soils
PIMPINELLA SAXIFRAGA	Burnet-saxifrage	Common; calcareous grassland & churchyards
Pinguicula vulgaris *	Butterwort	Pre-1926 (near Winslow); bogs
Pinus contorta	Lodgepole Pine	Very rare; plantations; *e.g.* Hollywell Plantation
Pinus nigra ssp. laricio	Corsican Pine	Scarce; plantations
Pinus nigra ssp. nigra	Austrian Pine	Scarce; plantations
Pinus pinaster ssp. pinaster *	Maritime Pine	Pre-1926; plantations
Pinus pinea *	Stone Pine	1944 (Brands Hill Park); parkland
Pinus ponderosa	Western Yellow-pine	Very rare; parkland; *e.g.* Tiddenfoot Water Park
Pinus sylvestris	Scots Pine	Common; plantations & heaths; frequently self-sown
Pinus wallichiana *	Bhutan Pine	1944 (Brands Hill Park); parkland
Pisum sativum	Garden Pea	Pre-1986 (Gerrards Cross); waste places
Pistia stratiotes	Water Lettuce	1996 (Buckingham); pond
Plagiobothrys scouleri	White Forget-me-not	1989 (Salden); reseeded grassland
Plantago arenaria *	Branched Plantain	Pre-1926 (Eton & Langley); waste places
PLANTAGO CORONOPUS	Buck's-horn Plantain	Rare; short turf on sandy soils
Plantago lagopus *	Hare's-foot Plantain	Pre-1926 (near Uxbridge); waste places
PLANTAGO LANCEOLATA	Ribwort Plantain	Very common; grassland & waste places
PLANTAGO MAJOR SSP. INTERMEDIA		Uncommon; damp places; possibly under-recorded
PLANTAGO MAJOR SSP. MAJOR	Greater Plantain	Very common; disturbed ground, waysides & grassy places
PLANTAGO MEDIA	Hoary Plantain	Common; dry grassy places on calcareous soils; often in churchyards
Platanthera bifolia	Lesser Butterfly-orchid	VU; 1984 (Ludgershall); woods
PLATANTHERA CHLORANTHA	Greater Butterfly-orchid	NT; very rare; woods; especially on heavy soils; *e.g.* Claydon woods
Platanus x hispanica	London Plane	Rare; planted in urban areas; rarely self-sown
Pleioblastus chino	Chino Bamboo	Very rare; RAF Sailing Club, Medmenham
POA ANGUSTIFOLIA	Narrow-leaved Meadow-grass	Rare; grassy places; possibly under-recorded
POA ANNUA	Annual Meadow-grass	Very common; grassy & bare places; often a weed
Poa chaixii	Broad-leaved Meadow-grass	1986 (Cliveden); woodland
POA COMPRESSA	Flattened Meadow-grass	Rare; dry bare places & walls
POA HUMILIS	Spreading Meadow-grass	Rare; on light soils or by water; possibly under-recorded
POA NEMORALIS	Wood Meadow-grass	Locally common; woods
Poa palustris *	Swamp Meadow-grass	1914 (Slough); rubbish heap
POA PRATENSIS	Smooth Meadow-grass	Very common; grassland & woodland rides
POA TRIVIALIS	Rough Meadow-grass	Very common; grassland, woodland rides & roadsides
Polemonium caeruleum	Jacob's-ladder	Very rare; roadsides; *e.g.* Cheddington
Polemonium foliosissimum	Leafy Jacob's-ladder	Very rare; railway bank at Windmill Plantation
Polycarpon tetraphyllum	Four-leaved Allseed	1990 (Boveney); fallow field
POLYGALA CALCAREA	Chalk Milkwort	Very rare; calcareous grassland; only known from Aston Clinton ragpits
POLYGALA SERPYLLIFOLIA	Heath Milkwort	1988 (Frieth); acid grassland
Polygala vulgaris ssp. collina *		Pre-1926; calcareous grassland; probably overlooked
POLYGALA VULGARIS SSP. VULGARIS	Common Milkwort	Locally common; dry open areas, especially on chalk
POLYGALA VULGARIS X CALCAREA		1994 (Aston Clinton ragpits); calcareous grassland

POLYGONATUM MULTIFLORUM	Solomon's-seal	Very rare; woods; *e.g.* Great Kingshill
Polygonatum x hybridum	Garden Solomon's-seal	Rare; woods, hedges & waste places
Polygonum arenarium ssp. pulchellum	Lesser Red-knotgrass	1972 (Iver); rubbish tip
POLYGONUM ARENASTRUM	Equal-leaved Knotgrass	Scarce; cornfields & waste places
POLYGONUM AVICULARE	Knotgrass	Very common; arable fields & rough, trampled places
Polygonum graminifolium *		Pre-1926 (Langley & Iver); waste places
POLYGONUM RURIVAGUM	Cornfield Knotgrass	Very rare; arable fields on light soils
POLYPODIUM INTERJECTUM	Intermediate Polypody	Common; walls, hedgebanks & churchyards
POLYPODIUM VULGARE	Polypody	Common; walls, hedgebanks & churchyards
POLYPODIUM X MANTONIAE		Very rare; wall at Hughenden Valley
Polypogon monspeliensis	Annual Beard-grass	2002 (Hedgerley); waste places
Polypogon viridis	Water Bent	1993 (Booker); between paving in garden centre
POLYSTICHUM ACULEATUM	Hard Shield-fern	Very rare; woods; especially in the south
POLYSTICHUM SETIFERUM	Soft Shield-fern	Rare; dry woods & shady hedges; often introduced
Pontederia cordata	Pickerelweed	Very rare; in water; *e.g.* Loughton Brook
Populus alba	White Poplar	Uncommon; riversides & parkland
Populus balsamifera *	Eastern Balsam-poplar	Pre-1926; parkland
Populus nigra 'Italica'	Lombardy-poplar	Common; parkland & plantations
POPULUS NIGRA SSP. BETULIFOLIA	Black-poplar	Locally common; ditches & hedgerows; especially in Aylesbury Vale
POPULUS TREMULA	Aspen	Uncommon; wet woods
Populus trichocarpa	Western Balsam-poplar	Very rare; planted by water; *e.g.* Jubilee Pit
Populus x canadensis	Black-Italian Poplar	Common; plantations, roadsides & by water
Populus x canescens	Grey Poplar	Uncommon; hedgerows
Populus x jackii	Balm-of-Gilead	Very rare; wet woods; *e.g.* Danesfield
Portulaca oleracea *	Common Purslane	Pre-1926 (Taplow); waste ground
Potamogeton alpinus *	Red Pondweed	Pre-1926 (Great Brickhill & Water Eaton); ponds
POTAMOGETON BERCHTOLDII	Small Pondweed	Rare; canal & lakes
POTAMOGETON COMPRESSUS	Grass-wrack Pondweed	EN; very rare; canal & rivers; *e.g.* Startop's End
POTAMOGETON CRISPUS	Curled Pondweed	Scarce; ponds, streams, rivers & canals
POTAMOGETON FRIESII	Flat-stalked Pondweed	NT; rare; canal & ditch; *e.g.* Stanton Low
POTAMOGETON LUCENS	Shining Pondweed	Rare; rivers & canals
POTAMOGETON NATANS	Broad-leaved Pondweed	Uncommon; ponds, lakes & ditches
Potamogeton nodosus	Loddon Pondweed	VU; 1985 (Taplow & Dorney); river
Potamogeton obtusifolius	Blunt-leaved Pondweed	1973 (Cub Pond); ponds
POTAMOGETON PECTINATUS	Fennel Pondweed	Uncommon; canal & lakes
POTAMOGETON PERFOLIATUS	Perfoliate Pondweed	Scarce; canal & rivers
POTAMOGETON POLYGONIFOLIUS	Bog Pondweed	Very rare; bog pools; *e.g.* Burnham Beeches & Black Park
POTAMOGETON PRAELONGUS	Long-stalked Pondweed	NT; 1991 (Horton Wharf); rivers & canal
POTAMOGETON PUSILLUS	Lesser Pondweed	Rare; ponds, lakes & canal
POTAMOGETON TRICHOIDES	Hairlike Pondweed	Very rare; pond at Olney Park
Potamogeton x cooperi *	Cooper's Pondweed	1945 (Startop's End); canal
Potamogeton x salicifolius	Willow-leaved Pondweed	Pre-1985 (Aston Clinton); canal
POTENTILLA ANGLICA	Trailing Tormentil	Scarce; woodland rides & grassy places; not on calcareous soils
POTENTILLA ANSERINA	Silverweed	Common; damp grassland, roadsides; woodland rides & by water
Potentilla argentea	Hoary Cinquefoil	NT; 1978 (East Burnham Common); dry sandy places
POTENTILLA ERECTA SSP. ERECTA	Tormentil	Uncommon; acid grassland & heaths
Potentilla fruticosa	Shrubby Cinquefoil	NT; 1994 (Bierton Pits); waste places
Potentilla hirta *	Hairy Cinquefoil	Pre-1926 (Iver); waste places
Potentilla intermedia	Russian Cinquefoil	Pre-1985 (Buckingham & Shenley Brook End); waste places
Potentilla nepalensis	Nepal Cinquefoil	1998 (Stoke Hammond); roadside

Potentilla neumanniana	Spring Cinquefoil	Very rare; only known from Prestwood churchyard
Potentilla norvegica	Ternate-leaved Cinquefoil	1998 (Caldecotte Lake); grassy & waste places
Potentilla recta	Sulphur Cinquefoil	Very rare; waste places & scrub
POTENTILLA REPTANS	Creeping Cinquefoil	Very common; grassy places & lawns
POTENTILLA STERILIS	Barren Strawberry	Common; woods, hedgerows & walls
Potentilla x italica		1984 (Dropmore); churchyard
POTENTILLA X MIXTA		2001 (Finemere Wood); grassy ride
PRIMULA ELATIOR	Oxlip	NT; very rare; woods; only known from near the Herts border; planted elsewhere
PRIMULA VERIS	Cowslip	Common; dry grassland, roadsides, downland & railway banks
PRIMULA VULGARIS	Primrose	Common; woods & hedgebanks
Primula x digenea		Very rare; plantation at Woolstone; arose naturally from planted parents
PRIMULA X POLYANTHA	False Oxlip	Scarce; woods, especially on open rides
Prunella laciniata	Cut-leaved Selfheal	*c.*1975 (Brill); railway banks & calcareous grassland
PRUNELLA VULGARIS	Selfheal	Very common; short grassland, woodland rides & churchyards
Prunella x intermedia		Pre-1975 (unlocalised); chalk grassland
PRUNUS AVIUM	Wild Cherry	Locally common; woods; especially on The Chilterns
Prunus cerasifera	Cherry Plum	Common; hedges; especially in urban areas
PRUNUS CERASUS	Dwarf Cherry	1978 (Cowcroft); woods & hedges
Prunus domestica ssp. domestica	Plum	Frequent; hedges
PRUNUS DOMESTICA SSP. INSITITIA	Bullace	Scarce; hedges
Prunus domestica ssp. italica	Greengage	Rare; hedges
Prunus dulcis	Almond	2005 (Four Ashes); parkland & cemetery
Prunus laurocerasus	Cherry Laurel	Common; woods, urban plantings & churchyards; frequently self-sown
Prunus lusitanica	Portuguese Laurel	Rare; woods & churchyards
Prunus mahaleb	St Lucie Cherry	1987 (Bow Wood); woodland
Prunus padus	Bird Cherry	Rare; hedges & urban specimen trees
Prunus serotina	Rum Cherry	1998 (Pancake & Hockeridge Woods); woodland
PRUNUS SPINOSA	Blackthorn	Very common; hedgerows & woods
Pseudofumaria alba	Pale Corydalis	Very rare; walls & under trees; *e.g.* Woburn Sands
Pseudofumaria lutea	Yellow Corydalis	Common; walls near habitation
Pseudosasa japonica	Arrow Bamboo	1997 (New Bradwell); pathsides & roadsides
Pseudotsuga menziesii	Douglas Fir	Very rare; plantations
PTERIDIUM AQUILINUM SSP. AQUILINUM	Bracken	Common; woods, heaths & railway banks on acid soils
Pteris cretica	Ribbon Fern	Very rare; wall at Denham
Puccinellia distans ssp. distans	Reflexed Saltmarsh-grass	1996 (Thorney); roadside
PULICARIA DYSENTERICA	Common Fleabane	Uncommon; marshy grassland, especially on clay soils
Pulicaria vulgaris *	Small Fleabane	1952 (Littleworth Common); pond margin
Pulmonaria officinalis	Lungwort	Rare; hedges & roadsides near habitation
Pulmonaria rubra	Red Lungwort	1998 (North Crawley); churchyard
PULSATILLA VULGARIS	Pasque-flower	VU; very rare; calcareous grassland in the Ivinghoe Hills
Pyracantha coccinea	Firethorn	Rare; hedges & waste places
PYROLA MINOR	Common Wintergreen	Very rare; woodland; decreasing
Pyrus communis	Pear	Rare; hedges
Pyrus pyraster *	Wild Pear	1905 (near Kimble); hedges
Quercus canariensis	Algerian Oak	1970 (Wotton Park); parkland
Quercus cerris	Turkey Oak	Rare; plantations & specimen trees
Quercus dentata	Daimio Oak	1977 (Church Wood); woodland
Quercus georgiana *	Georgia Oak	1951 (New Wavendon Heath); woodland
Quercus ilex	Evergreen Oak	Rare; plantations, parkland & churchyards
Quercus palustris	Pin Oak	1998 (Hockeridge Wood); woodland
QUERCUS PETRAEA	Sessile Oak	Rare; woods & heaths on acid soil
QUERCUS ROBUR	Pedunculate Oak	Very common; woods & hedgerows
Quercus rubra	Red Oak	Rare; woods & specimen trees
Quercus suber	Cork Oak	1997 (Wotton Park); parkland
Quercus x crenata	Lucombe Oak	Very rare; parkland at Prestwood Park
QUERCUS X ROSACEA		Rare; woods; probably under-recorded
Radiola linoides	Allseed	1972 (Stoke Common); commons in the south
RANUNCULUS ACRIS	Meadow Buttercup	Very common; especially in old meadows & other grasslands
RANUNCULUS AQUATILIS	Common Water-crowfoot	Rare; ponds & slow-flowing water

RANUNCULUS ARVENSIS	Corn Buttercup	CR; very rare; arable fields & waste places; formerly more common
RANUNCULUS AURICOMUS	Goldilocks Buttercup	Uncommon; woodland & shady churchyards
RANUNCULUS BULBOSUS	Bulbous Buttercup	Common; old meadows, churchyards & pastures
RANUNCULUS CIRCINATUS	Fan-leaved Water-crowfoot	Very rare; ponds & slow-flowing water; *e.g.* West Wycombe
RANUNCULUS FICARIA SSP. BULBILIFER		Common; woods, hedges, banks & churchyards
RANUNCULUS FICARIA SSP. FICARIA	Lesser Celandine	Common; woods, hedges, banks & churchyards
Ranunculus ficaria ssp. ficariiformis	Giant Celandine	1997 (Farnham Common); hedge
RANUNCULUS FLAMMULA SSP. FLAMMULA	Lesser Spearwort	Scarce; ponds, ditches & bogs; not on calcareous soils
RANUNCULUS FLUITANS	River Water-crowfoot	Rare; larger rivers
RANUNCULUS HEDERACEUS	Ivy-leaved Crowfoot	Very rare; bogs & spring-fed streams on acid soils
RANUNCULUS LINGUA	Greater Spearwort	Rare; ponds, lakes & rivers; probably extinct as a native
Ranunculus marginatus	St Martin's Buttercup	1994 (Hazeley Wood); imported in wild flower seed mix
RANUNCULUS PARVIFLORUS	Small-flowered Buttercup	Very rare; short sandy turf in south
RANUNCULUS PELTATUS	Pond Water-crowfoot	Uncommon; ponds & slow-flowing water
RANUNCULUS PENICILLATUS SSP. PSEUDOFLUITANS	Stream Water-crowfoot	Rare; rivers & streams
RANUNCULUS REPENS	Creeping Buttercup	Very common; grassland, rough places, gardens & pond margins
RANUNCULUS SARDOUS	Hairy Buttercup	1996 (Colnbrook); pasture
RANUNCULUS SCELERATUS	Celery-leaved Buttercup	Common; pond margins & cattle-poached river banks
RANUNCULUS TRICHOPHYLLUS	Thread-leaved Water-crowfoot	Rare; ponds & wet ditches
RAPHANUS RAPHANISTRUM SSP. RAPHANISTRUM	Wild Radish	Uncommon; arable fields & waste places; especially on light soils
Raphanus sativus	Garden Radish	1995 (Emberton); garden weed; from birdseed origin
Rapistrum perenne	Steppe Cabbage	Pre-1926 (Slough & Iver); waste places
Rapistrum rugosum ssp. linnaeanum	Bastard Cabbage	1998 (High Wycombe); waste places
Reseda alba *	White Mignonette	1903 (Marsh Gibbon); railway bank
RESEDA LUTEA	Wild Mignonette	Uncommon; open ground on calcareous soils
RESEDA LUTEOLA	Weld	Uncommon; railway ballast & waste ground; especially on heavy soils
Reseda odorata *	Garden Mignonette	Pre-1926 (Slough & Eton); tips & waste places
Rhagadiolus stellaris	Star Hawkbit	1970 (Beaconsfield); waste places
RHAMNUS CATHARTICA	Buckthorn	Uncommon; hedges & wood margins; especially on calcareous soils
Rheum x hybridum	Rhubarb	Rare; often persistant on old allotments
Rhinanthus angustifolius *	Greater Yellow-rattle	1901 (Edlesborough); cultivated ground
Rhinanthus minor ssp. calcareus *		1831 (Missenden); grassland
RHINANTHUS MINOR SSP. MINOR	Yellow-rattle	Uncommon; old meadows & pastures
Rhinanthus minor ssp. stenophyllus *		Pre-1926; pastures
Rhododendron luteum	Yellow Azalea	Rare; woodland on acid soils; *e.g.* Oak Wood & Burnham Beeches
Rhododendron ponticum	Rhododendron	Locally common; woods on acid soils
Rhodotypos scandens	Black Jetbead	1998 (Central Milton Keynes); planted & rarely self-sown
Rhus typhina	Stag's-horn Sumach	Uncommon; roadsides & waste places; spreading by suckers
Rhynchospora alba *	White Beak-sedge	1959 (Burnham Beeches); wet heaths
Ribes alpinum	Mountain Currant	1988 (Barn Wood); woodland
Ribes nigrum	Black Currant	Rare; moist woods & waste places; especially in the south
Ribes rubrum	Red Currant	Uncommon; woods, shaded streambanks & waste places
Ribes sanguineum	Flowering Currant	Rare; walls & hedges
Ribes uva-crispa	Gooseberry	Uncommon; woods, hedgerows & waste places
Ricinus communis *	Castor-oil-plant	Pre-1926 (Iver); waste places
Ridolfia segetum	False Fennel	1985 (Amersham); waste places
Robinia pseudoacacia	False-acacia	Rare; woods & waste places; occasionally self-sown
Roemeria hybrida *	Violet Horned-poppy	1928 (Hambledon); waste places
RORIPPA AMPHIBIA	Great Yellow-cress	Scarce; river banks
RORIPPA MICROPHYLLA	Narrow-fruited Water-cress	Rare; ponds, streams & ditches
RORIPPA NASTURTIUM-AQUATICUM	Water-cress	Common; ponds, streams & ditches
Rorippa palustris ssp. hispida	Hispid Yellow-cress	1989 (Salden); introduced with grass seed

RORIPPA PALUSTRIS **SSP. PALUSTRIS**	Marsh Yellow-cress	Scarce; bare ground by ponds & ditches
RORIPPA SYLVESTRIS	Creeping Yellow-cress	Rare; garden weed or on walls
RORIPPA X STERILIS	Brown Water-cress	Rare; ponds, streams & ditches
Rosa agrestis *	Small-leaved Sweet-briar	NT; RDB; 1906 (Pheasants); bushy places on chalk; possibly overlooked
ROSA ARVENSIS	Field-rose	Very common; woodland margins & hedges
ROSA CANINA	Dog-rose	Very common; woods, hedgerows & scrub

DOG ROSE

Rosa ferruginea	Red-leaved Rose	Very rare; frequently planted, occasionally bird-sown
Rosa 'Hollandica'	Dutch Rose	1999 (Cholesbury); hedge
ROSA MICRANTHA	Small-flowered Sweet-briar	Uncommon; scrub in the north & on The Chilterns
Rosa multiflora	Many-flowered Rose	Rare; hedgerows
Rosa 'Nozomi'		c.1998 (Bradwell); roadside
Rosa obtusifolia *	Round-leaved Dog-rose	1906 (Whaddon); hedges & scrub; probably overlooked
ROSA RUBIGINOSA	Sweet-briar	Rare; hedges & scrub on the chalk; probably overlooked
Rosa rugosa	Japanese Rose	Rare; railway banks & waste places; e.g. Bletchley
Rosa spinosissima	Burnet Rose	Rare; hedges & churchyards
ROSA STYLOSA	Short-styled Field-rose	1970 (Brill); hedges
Rosa tomentosa *	Harsh Downy-rose	1912 (Princes Risborough); hedges
Rosa virginiana	Virginian Rose	1997 (Linford Wood); woodland
Rosa xanthina	Father Hugo's Rose	1996 (Hughenden); old allotments
Rosa x andegavensis		1997 (Bovingdon Green); hedges
ROSA X DUMALIS		1999 (Lenborough Wood); woods & hedges
Rosa x dumetorum *		1897 (Northall); hedges
ROSA X IRREGULARIS		1992 (Salcey Forest); woods & hedges; overlooked
Rosa x nitidula *		1882 (Wraysbury); hedges or scrub
ROSA X SCABRIUSCULA		1999 (Ivinghoe Bridge); hedges
Rosmarinus officinalis	Rosemary	1998 (Downs Barn); pavement cracks
RUBUS ACCLIVITATUM		1992; wood margins
Rubus adamsii *		1964; wood & heath margins; hedges
Rubus adspersus *		1952; heaths in the north & south
Rubus albionis *		1963; heathy places
RUBUS AMPLIFICATUS		1993; wood borders & hedges
Rubus armeniacus	Giant Blackberry	1986; garden escape; increasing rapidly in waste places
RUBUS ARMIPOTENS		1993; wood margins & hedges in heathy areas
Rubus atrebatum *		1963; hedges & heaths
Rubus bercheriensis		1976; wood margins, hedges & heaths
Rubus bloxamianus *		1963; wood margins, hedges & heaths
RUBUS BRITANNICUS		1993; woods & wood borders
RUBUS CAESIUS	Dewberry	Common; woody areas on damp calcareous soils
Rubus cantianus *		1939; wood margins, hedges & heaths
Rubus cardiophyllus		1976; throughout; especially hedges on heaths
Rubus cinerosus *		1899; wood borders & heaths
Rubus cissburiensis *		1962; heathy areas in south
Rubus cockburnianus	White-stemmed Bramble	Rare; waste places; e.g. Wolverton
Rubus conjungens		1974; widespread; hedges
RUBUS CRINIGER		1993; hedges & heaths in the south
RUBUS CURVISPINOSUS		1993 (Stoke Common); woods & heath borders
Rubus dasyphyllus		1976; woods etc. in heathy places
Rubus decussatus *		1964; wood margins & heaths
Rubus echinatoides		1976; throughout; woods, hedges & heaths
Rubus echinatus		1976; throughout; open woods, hedges & heath margins

Rubus egregius		1974; open woods & heaths
RUBUS EURYANTHEMUS		1993; oak/birch woods; heaths
Rubus fissus *		1961; heathy woods
Rubus flexuosus		1976; woods in the north & south
Rubus fuscus		1976; woods in the south
Rubus glareosus *		1925; hedges in the south
Rubus hylonomus *		1963; woods & heaths
Rubus hylophilus *		1963; woods & heath margins
RUBUS IDAEUS	Raspberry	Frequent; woods, mostly on heathy or light sandy soil
Rubus insectifolius *		1963; heathy areas
Rubus laciniatus	Cut-leaved Bramble	Rare; garden escape; waste places
Rubus largificus *		1922; wood borders, hedges & heaths
RUBUS LEIGHTONII		1993; hedges
Rubus leptadenes *		1968 (Angling Spring & Lee Clump); very rare
RUBUS LEUCOSTACHYS		1993; widespread; woods & hedges
Rubus lindebergii		1986; woods, off the acid soils
Rubus lindleianus		1974; hedges on heaths
Rubus macrophyllus *		1963; widespread; woods, hedges & heaths
Rubus milfordensis *		1948; wood borders & heaths
Rubus moylei *		1948; woods & hedge
Rubus murrayi *		1963; woods in the north & south
Rubus nemoralis *		1964; woods & heaths
Rubus nemorosus *		1934; rare; hedge
Rubus nessensis *		1963; heaths
Rubus nitidiformis		1986; rare; woods
RUBUS PEDEMONTANUS		1992; damp woods
Rubus phaeocarpus *		1963; wood borders & heaths
Rubus platyacanthus *		1963; wood borders hedges & heaths
Rubus plicatus		1974; dry heaths in the north

BRAMBLE

Rubus poliodes *		1964; wood margins, hedges & heaths
Rubus polyanthemus		1974; widespread; hedges & heaths
Rubus pruinosus		1976; widespread; woods & hedges
Rubus pyramidalis *		1963; woods & heathy places in the north & south
RUBUS RADULA		1992; throughout; woods & hedges
Rubus raduloides *		1951; woods & hedges on less acid soils
Rubus rudis		1976; open woods in the north & south
Rubus rufescens		1976; woods in the south especially on less acid soils
Rubus scaber *		1948; woods on light soils in the south
Rubus sciocharis		1974; wood borders & hedges
Rubus sprengelii *		1963; heaths in the north & south
RUBUS SUBINERMOIDES		1992; woods & shady hedges
Rubus subtercanens *		1948; wood borders & hedge
RUBUS SURREJANUS		1993; wood borders & heaths; hedges
RUBUS TRICHODES		1993; woods
RUBUS TUBERCULATUS		1993; railway banks & hedge
RUBUS ULMIFOLIUS		Common; hedges especially on calcareous soils
Rubus vestitus		1976; widespread & common
Rubus watsonii		1977; woods, hedges & heaths
Rubus wedgwoodiae *		1901; wood borders & heaths
RUBUS WIRRALENSIS		1993; woods & hedges on damp clay
Rudbeckia fulgida	Perennial Coneflower	2004 (Moor Wood); woodland
Rudbeckia hirta	Black-eyed-Susan	1997 (Chesham Moor); waste places
RUMEX ACETOSA **SSP. ACETOSA**	Common Sorrel	Very common; grassland, especially on calcareous soils

Species	Common Name	Notes
RUMEX ACETOSELLA	Sheep's Sorrel	Uncommon; light acid soils; both sspp. *acetosella* & *pyrenaicus* occur but distributions not known
RUMEX CONGLOMERATUS	Clustered Dock	Common; damp places; especially by water
RUMEX CRISPUS **SSP. CRISPUS**	Curled Dock	Very common; grassland & waste places
Rumex cristatus	Greek Dock	Very rare; waste ground at Handy Cross
RUMEX HYDROLAPATHUM	Water Dock	Uncommon; canal, river & pond banks
RUMEX MARITIMUS	Golden Dock	Very rare; margins of lakes & ditche
RUMEX OBTUSIFOLIUS	Broad-leaved Dock	Very common; bare ground & rough grassy places
RUMEX PALUSTRIS	Marsh Dock	1992 (Sutton); dry ditch
RUMEX PULCHER **SSP. PULCHER**	Fiddle Dock	Rare; churchyards & dry open grassland; especially in the north
RUMEX SANGUINEUS	Wood Dock	Common; woods & sheltered hedgerows
Rumex x abortivus *		1907 (Slough)
Rumex x dufftii *		Pre-1970 (unlocalised); waste places
RUMEX X PRATENSIS	Meadow Dock	Uncommon; grassy places
Rumex x sagorskii *		Pre-1970 (unlocalised); waste places
RUMEX X WEBERI		1986 (Wraysbury); roadside
RUSCUS ACULEATUS	Butcher's-broom	Rare; woods in the south
Ruscus hypoglossum	Spineless Butcher's-broom	1995 (Bow Brickhill); churchyard
Ruta graveolens	Rue	1995 (Longwick)
SAGINA APETALA **SSP. APETALA**	Ciliate Pearlwort	Rare; open gravelly or sandy places
SAGINA APETALA **SSP. ERECTA**	Annual Pearlwort	Uncommon; open gravelly or sandy places
Sagina nodosa *	Knotted Pearlwort	1902 (Chesham Moor); wet grassland
SAGINA PROCUMBENS	Procumbent Pearlwort	Common; damp grassland, pavements & gardens
Sagina subulata *	Heath Pearlwort	Pre-1926; dry sandy heaths in the south
SAGITTARIA SAGITTIFOLIA	Arrowhead	Uncommon; rivers & canals
SALIX ALBA	White Willow	Uncommon; by rivers & ditches
SALIX AURITA	Eared Willow	Very rare; heaths & damp woods
SALIX CAPREA **SSP. CAPREA**	Goat Willow	Common; woods, hedges & by water
SALIX CINEREA **SSP. CINEREA**	Grey Willow	1996 (Wraysbury); marshes
SALIX CINEREA **SSP. OLEIFOLIA**	Rusty Willow	Common; woods, hedges & by water
Salix daphnoides	European Violet-willow	Very rare; planted in urban areas & by water
Salix elaeagnos	Olive Willow	Very rare; frequently planted in urban areas or by water; rarely self-sown
SALIX FRAGILIS	Crack-willow	Common; wet woodland edges & by water
Salix pentandra	Bay Willow	1986 (Marlow); planted by water
SALIX PURPUREA	Purple Willow	Rare; stream & river banks, especially in the south
SALIX REPENS	Creeping Willow	Very rare; heaths; *e.g.* Stoke Common
SALIX TRIANDRA	Almond Willow	Rare; streambanks & hedges; probably under-recorded
Salix triandra x purpurea *		Pre-1926 (Aston Ferry); riversid
SALIX VIMINALIS	Osier	Common; by water & in wet wood
Salix x ambigua *		Pre-1926; heathland in the south; *e.g.* Fulmer
Salix x calodendron	Holme Willow	Very rare; planted by lake at New Bradwell
Salix x capreola *		Pre-1926 (Winslow & near Grendon Underwood)
Salix x ehrhartiana	Ehrhart's Willow	Very rare; osier bed at Emberton Park
Salix x forbyana	Fine Osier	Rare; hedgerows
Salix x fruticosa *	Shrubby Osier	Pre-1926 (Winslow
Salix x holosericea (S. smithiana)	Silky-leaved Osier	1995 (Loughton); hedges & by water
Salix x meyeriana	Shiny-leaved Willow	Very rare; rough places & hedges; *e.g.* Bletchley
Salix x mollissima	Sharp-stipuled Willow	Very rare; scrub at Joan's Piece
Salix x multinervis *		Pre-1926; heathy places
SALIX X REICHARDTII		Scarce; hedges & by water
Salix x rubens	Hybrid Crack-willow	Rare; by water; *e.g.* Water Stratford
Salix x rubra *	Green-leaved Willow	Pre-1926 (Taplow); by water
Salix x sepulcralis	Weeping Willow	Frequent; planted by water & in parkland
Salix x smithiana (S. sericans)	Broad-leaved Osier	1995 (Great Horwood); hedges
Salix x stipularis *	Eared Osier	1789 (Denham)
Salsola kali ssp. iberica *	Spineless Saltwort	1958 (Wooburn); paper mill waste
Salvia nemorosa *	Balkan Clary	Pre-1926 (Slough); rubbish heaps
Salvia patens	Gentian Sage	1989 (Milton Keynes Village); roadside
SALVIA PRATENSIS	Meadow Clary	NT; very rare; calcareous grassland; *e.g.* Dancersend

Salvia sclarea	Clary	1997 (Great Missenden); walls & rubbish heaps
SALVIA VERBENACA	Wild Clary	Very rare; walls & dry grassland; much decreased
Salvia verticillata *	Whorled Clary	1938 (Cheddington); railway banks & waste places
Salvia x sylvestris *		Pre-1926 (near Leighton Buzzard); waste places
Salvinia auriculata agg.	Water Spangles	2003 (Chesham Bois Common); pond
Sambucus canadensis	Canadian Elder	Very rare; planted in urban areas; rarely self-sown
SAMBUCUS EBULUS	Dwarf Elder	Very rare; hedgerows; *e.g.* Thornborough
SAMBUCUS NIGRA	Elder	Very common; woods, hedges & waste places
SAMOLUS VALERANDI	Brookweed	Very rare; wet grassland by water at Linford Pits
SANGUISORBA MINOR SSP. MINOR	Salad Burnet	Common; calcareous grassland

SALAD BURNET

Sanguisorba minor ssp. muricata	Fodder Burnet	Very rare; dry places; much decreased
SANGUISORBA OFFICINALIS	Great Burnet	Scarce; wet meadows in the north & west
SANICULA EUROPAEA	Sanicle	Uncommon; woodland
Saponaria officinalis	Soapwort	Rare; waste places & hedges
Sasa palmata	Broad-leaved Bamboo	Very rare; parkland at Black Park
Sasa veitchii	Veitch's Bamboo	Very rare; lakeside at Cliveden
Sasaella ramosa	Hairy Bamboo	1971 (Cliveden); parkland
Satureja montana	Winter Savory	Very rare; wall at Whitchurch
Saxifraga cymbalaria	Celandine Saxifrage	Very rare; garden weed at Willen
SAXIFRAGA GRANULATA	Meadow Saxifrage	Rare; churchyards & meadows; *e.g.* Tingewick
SAXIFRAGA TRIDACTYLITES	Rue-leaved Saxifrage	Scarce; walls & sandy ground
Saxifraga x arendsii Group	Garden Mossy-saxifrage	1998 (Beaconsfield); cemetery
Saxifraga x urbium	Londonpride	Very rare; walls at Stony Stratford
Scabiosa atropurpurea	Sweet Scabious	1997 (College Lake); chalk grassland
SCABIOSA COLUMBARIA	Small Scabious	Uncommon; chalk grassland
SCANDIX PECTEN-VENERIS	Shepherd's-needle	CR; very rare; arable fields & gardens; occasionally from birdseed
SCHOENOPLECTUS LACUSTRIS	Common Club-rush	Uncommon; in rivers, canals & rarely ponds
SCHOENOPLECTUS TABERNAEMONTANI	Grey Club-rush	Very rare; ponds & lakes; sometimes introduced
Schoenus nigricans *	Black Bog-rush	1899 (near Winslow); marshes
Scilla bifolia	Alpine Squill	1992 (Windmill Plantation); woodland
Scilla siberica	Siberian Squill	1992 (Bragenham); roadside
Scirpus sylvaticus	Wood Club-rush	Very rare; marshy places & wet woods; *e.g.* Burnham Beeches
SCLERANTHUS ANNUUS SSP. ANNUUS	Annual Knawel	EN; very rare; sandy fields & roadsides
Scolymus hispanicus	Golden Thistle	Pre-1985 (Wotton Underwood); waste places
Scorpiurus muricatus	Caterpillar-plant	1983 (Stewkley); garden; originating from birdseed
Scorpiurus vermicularis	Sheep's-tongue	1970 (Weedon); roadside
Scorzonera hispanica	Scorzonera	1996 (Heelands); waste ground
SCROPHULARIA AURICULATA	Water Figwort	Common; by water & in wet woods
SCROPHULARIA NODOSA	Common Figwort	Common; woods & hedges
Scrophularia vernalis	Yellow Figwort	Very rare; woodland at Cliveden
SCUTELLARIA GALERICULATA	Skullcap	Uncommon; by rivers & canals
SCUTELLARIA MINOR	Lesser Skullcap	Very rare; wet places on acid commons in the south
Secale cereale	Rye	1996 (Water Stratford); waste places & a relic of cultivation
Securigera varia	Crown Vetch	1989 (Boveney); riverbanks & waste places
Sedum 'Herbstfreude'	Autumn Joy Stonecrop	1997 (Heelands); hedge
SEDUM ACRE	Biting Stonecrop	Uncommon; walls & dry open soil; often near habitation

Scientific name	Common name	Notes
Sedum aizoon	Orange Stonecrop	Very rare; abandoned bird-bath at Medmenham
Sedum album	White Stonecrop	Rare; walls; usually near habitation
Sedum cepaea	Pink Stonecrop	Pre-1926; banks in the south; *e.g.* Hedgerley & Denham
Sedum dasyphyllum	Thick-leaved Stonecrop	Very rare; walls & banks; *e.g.* Marsh Gibbon
Sedum fosterianum	Rock Stonecrop	Very rare; walls; *e.g.* Thornborough
Sedum hispanicum	Spanish Stonecrop	Very rare; gravel path in Marsh Gibbon churchyard
Sedum rupestre	Reflexed Stonecrop	Common; walls & waste places
Sedum sexangulare *	Tasteless Stonecrop	1864 (Danesfield); walls
Sedum spectabile	Butterfly Stonecrop	Very rare; waste places
Sedum spurium	Caucasian-stonecrop	Rare; walls; *e.g.* Loughton
SEDUM TELEPHIUM	Orpine	Rare; hedgerows & wood margins on chalk or sandy soils
Sempervivum arachnoideum	Cobweb House-leek	1998 (Lavendon Mill); walls
Sempervivum tectorum	House-leek	Rare; walls & rooftops; especially on limestone
SENECIO AQUATICUS	Marsh Ragwort	Rare; marshes & wet meadows
Senecio cineraria	Silver Ragwort	Very rare; waste places near habitation; *e.g.* Upper Hartwell
SENECIO ERUCIFOLIUS	Hoary Ragwort	Common; roadsides, grassland & railway banks
Senecio fluviatilis	Broad-leaved Ragwort	1970 (Wilstone Reservoir); waterside
SENECIO JACOBAEA	Common Ragwort	Very common; grassland & waste places
Senecio squalidus	Oxford Ragwort	Uncommon; walls, waste places & railway banks
SENECIO SYLVATICUS	Heath Groundsel	Rare; dry sandy soils in the Brickhills & the south
Senecio viscosus	Sticky Groundsel	Uncommon; waste places & railway ballast
SENECIO VULGARIS	Groundsel	Very common; cultivated & waste places
Senecio x albescens		1998 (Wendover); garden weed
Senecio x subnebrodensis	London Groundsel	1997 (Princes Risborough); waste places & disused railway
Sequoia sempervirens	Coastal Redwood	1988 (Common Wood); woodland
Sequoiadendron giganteum	Wellingtonia	Very rare; planted in woods & parkland; *e.g.* Broomhills Wood
Serratula tinctoria	Saw-wort	Rare; fens & wet grassland
Setaria italica	Foxtail Bristle-grass	Pre-1985 (Langley Park); rubbish heap
Setaria pumila	Yellow Bristle-grass	Rare; roadsides & waste places; *e.g.* Bletchley
Setaria verticillata *	Rough Bristle-grass	Pre-1926 (Slough & Iver); waste places
Setaria viridis	Green Bristle-grass	Scarce; cultivated fields & waste places; *e.g.* Burcott
SHERARDIA ARVENSIS	Field Madder	Uncommon; cultivated ground, short turf & waste places
SILAUM SILAUS	Pepper-saxifrage	Uncommon; wet calcareous meadows or fens
Silene armeria	Sweet-William Catchfly	Very rare; garden weed; *e.g.* Aston Abbotts
Silene dichotoma *	Forked Catchfly	Pre-1926; cultivated fields; *e.g.* near Lane End
SILENE DIOICA	Red Campion	Locally common; hedges & roadsides on light soils
SILENE GALLICA	Small-flowered Catchfly	EN; 1975 (Amersham); cultivated fields
SILENE LATIFOLIA SSP. ALBA	White Campion	Common; hedges, roadsides & grassy places
Silene muscipula *	Catchfly	1928 (Beaconsfield); waste places
SILENE NOCTIFLORA	Night-flowering Catchfly	VU; very rare; cultivated fields; *e.g.* Turweston
SILENE VULGARIS SSP. VULGARIS	Bladder Campion	Frequent; dry open grassland
Silene x hampeana	Pink Campion	Common; hedges, roadsides & grassy places
Silybum marianum	Milk Thistle	Rare; waste places & roadsides; rarely naturalised; *e.g.* Marlow
SINAPIS ALBA SSP. ALBA	White Mustard	Scarce; arable fields & waste places; mostly on chalk soils
SINAPIS ARVENSIS	Charlock	Common; arable fields & waste places
SISON AMOMUM	Stone Parsley	Uncommon; hedgerows
Sisymbrium altissimum	Tall Rocket	Very rare; grassy & waste places
Sisymbrium irio *	London-rocket	Pre-1926 (Eton & Slough); waste places
SISYMBRIUM OFFICINALE	Hedge Mustard	Very common; cultivated ground, roadsides & hedges
Sisymbrium orientale	Eastern Rocket	Rare; waste places in urban areas
Sisyrinchium striatum	Pale Yellow-eyed-grass	1999 (Central Milton Keynes); waste places
Sium latifolium	Greater Water-parsnip	EN; pre-1985 (Harleyford); rivers
Smyrnium olusatrum	Alexanders	Rare; roadsides & waste places; *e.g.* Denham
Smyrnium perfoliatum	Perfoliate Alexanders	1999 (Cliveden); parkland
SOLANUM DULCAMARA	Bittersweet	Very common; woods, hedges & watersides
SOLANUM NIGRUM SSP. NIGRUM	Black Nightshade	Frequent; arable fields & gardens
Solanum rostratum	Buffalo-bur	1996 (Stewkley); garden weed originating from birdseed
Solanum tuberosum	Potato	Rare; waste places & rubbish dumps
Soleirolia soleirolii	Mind-your-own-business	Uncommon; damp places, especially at bases of walls
Solidago canadensis	Canadian Goldenrod	Common; roadsides & waste places; increasing

Solidago gigantea ssp. serotina	Early Goldenrod	Rare; woods & waste places; *e.g.* Moor Wood
SOLIDAGO VIRGAUREA	Goldenrod	Very rare; heaths & woods on acid soils in the south
SONCHUS ARVENSIS	Perennial Sow-thistle	Common; roadsides, field margins & riversides
SONCHUS ASPER	Prickly Sow-thistle	Very common; arable fields, roadsides & waste places
SONCHUS OLERACEUS	Smooth Sow-thistle	Common; waste places; especially in urban areas
Sonchus palustris *	Marsh Sow-thistle	1909 (near Hell Coppice); damp hedgerow
Sorbaria sorbifolia	Sorbaria	1996 (Heelands); waste ground
SORBUS ARIA	Common Whitebeam	Locally common; woods & hedges; especially on The Chilterns
SORBUS AUCUPARIA	Rowan	Locally common; woods & heaths
Sorbus hybrida	Swedish Service-tree	1997 (Princes Risborough); seedling in a garden
Sorbus intermedia	Swedish Whitebeam	Uncommon; woods, walls & waste places; often planted
Sorbus latifolia	Broad-leaved Whitebeam	Very rare; carpark at Combe Hill
SORBUS TORMINALIS	Wild Service-tree	Very rare; woods; occasionally planted
SORBUS X THURINGIACA	Bastard Service-tree	Very rare; woodland; *e.g.* Burnham Beeches
Sorghum bicolor *	Great Millet	Pre-1926 (Iver); waste places
Sorghum halepense	Johnson-grass	1998 (Bucklandwharf); field
SPARGANIUM EMERSUM	Unbranched Bur-reed	Rare; rivers
SPARGANIUM ERECTUM	Branched Bur-reed	Frequent; in water; sspp. *erectum, microcarpum* & *neglectum* all occur but distributions not known
Spartium junceum	Spanish Broom	Very rare; planted on roadsides; rarely self-sown
SPERGULA ARVENSIS	Corn Spurrey	VU; rare; dry open places on sandy or gravelly soils
Spergularia marina	Lesser Sea-spurrey	Very rare; salted road verges
SPERGULARIA RUBRA	Sand Spurrey	Rare; dry open places on sandy or gravelly soils; especially in the south
Spinacia oleracea	Spinach	*c.*1992 (Downs Barn); pathside & rubbish heaps
Spiraea canescens	Himalayan Spiraea	1996 (Chalfont St Giles); churchyard
Spiraea douglasii ssp. douglasii	Steeple-bush	1996 (Widmer End); hedgerow
Spiraea japonica	Japanese Spiraea	1996 (Woburn Sands); pavement crack
Spiraea x billardii	Billard's Bridewort	1996 (Naphill); pondside
SPIRANTHES SPIRALIS	Autumn Lady's-tresses	NT; very rare; chalk grassland; much decreased
SPIRODELA POLYRHIZA	Greater Duckweed	Rare; ponds, rivers & canals; probably under-recorded
Spirodela punctata	Dotted Duckweed	Very rare; at present only two records as a Garden Centre weed
Stachys annua	Annual Woundwort	1997 (Heelands); garden weed & cultivated fields
STACHYS ARVENSIS	Field Woundwort	NT; very rare; cultivated fields in the south
Stachys byzantina	Lamb's-ear	1996 (Chalfont St Giles); churchyard
STACHYS OFFICINALIS	Betony	Uncommon; fens, wet grassland & woodland rides
STACHYS PALUSTRIS	Marsh Woundwort	Uncommon; by rivers, canals & ponds
STACHYS SYLVATICA	Hedge Woundwort	Very common; woods & hedgerows
STACHYS X AMBIGUA		Very rare; hedges; *e.g.* Springfield
Staphylea pinnata	Bladder-nut	Very rare; hedges; *e.g.* Bledlow
STELLARIA GRAMINEA	Lesser Stitchwort	Common; damp grassland & woodland rides
STELLARIA HOLOSTEA	Greater Stitchwort	Common; woods & hedgerows
STELLARIA MEDIA	Common Chickweed	Very common; cultivated land & open places in woods or grassland
Stellaria neglecta *	Greater Chickweed	Pre-1926 (near Burnham); hedgebanks & woods
STELLARIA PALLIDA	Lesser Chickweed	Very rare; sandy places & walls; now only known from Partridge Hill
STELLARIA PALUSTRIS	Marsh Stitchwort	VU; very rare; wet grassy places in the south; much decreased
STELLARIA ULIGINOSA	Bog Stitchwort	Uncommon; wet grassland & woodland rides on acid soils
Stratiotes aloides	Water-soldier	NT; very rare; ponds; *e.g.* Aylesbury; often planted
SUCCISA PRATENSIS	Devil's-bit Scabious	Uncommon; fens, heaths, wet grassland & chalk slopes
Symphoricarpos albus	Snowberry	Common; hedges, woods & parkland near habitation
Symphoricarpos x chenaultii	Hybrid Coralberry	Rare; much planted in urban areas; occasionally self-sown
Symphytum 'Hidcote Blue'	Hidcote Comfrey	1997 (Weedon); parkland
Symphytum grandiflorum	Creeping Comfrey	Rare; hedges & parkland near habitation
SYMPHYTUM OFFICINALE SSP. OFFICINALE	Common Comfrey	Uncommon; by rivers & canals, rarely in wet grassland
Symphytum orientale	White Comfrey	Scarce; roadsides & churchyards; especially near habitation
Symphytum tuberosum	Tuberous Comfrey	Very rare; woodland at Cliveden
Symphytum x uplandicum	Russian Comfrey	Uncommon; rough grassland, ditches & roadsides
Syringa vulgaris	Lilac	Uncommon; churchyards, hedges & parkland; often suckering
Tagetes patula	French Marigold	1995 (unlocalised); waste places
Tamarix gallica	Tamarisk	1983 (Chesham); waste places
TAMUS COMMUNIS	Black Bryony	Common; hedges & wood margins
TANACETUM PARTHENIUM	Feverfew	Common; hedges, walls & waste places

TANACETUM VULGARE	Tansy	Scarce; roadsides, railway banks & by rivers; sometimes planted
TARAXACUM ALATUM		Roadsides & waste places
TARAXACUM ANCISTROLOBUM		Roadsides & waste places
Taraxacum anglicum *		Seasonally wet grassland
Taraxacum argutum *		Especially on limestone soils
Taraxacum aurosulum *		Shady roadsides
TARAXACUM BRACHYGLOSSUM		Dry places
Taraxacum cophocentrum *		Wood margins, scrub & grassy places
TARAXACUM CORDATUM		Roadsides & waste places
TARAXACUM CROCEIFLORUM		Roadsides & waste places
Taraxacum cyanolepis *		Damp grassland
TARAXACUM ECKMANII		Roadsides & waste places
TARAXACUM EXPALLIDIFORME		Roadsides & waste places
Taraxacum falcatum		Dry grassland
Taraxacum faroense *		Seasonally wet grassland
Taraxacum fulviforme *		Dry calcareous grassland
TARAXACUM FULVUM		Light neutral or calcareous soils
Taraxacum gelertii *		Non-acid grassland
Taraxacum glauciniforme *		Light neutral or calcareous soils
TARAXACUM HAMATUM		Roadsides, waste places & gardens
TARAXACUM LACINIOSIFRONS		Roadsides
TARAXACUM LACISTOPHYLLUM		Light neutral or calcareous soils
TARAXACUM LATISECTUM		Fertile pastures
Taraxacum macrolobum *		Grassy places & roadsides
TARAXACUM MARKLUNDII		Roadsides & waste places
Taraxacum nordstedtii *		Wet meadows
Taraxacum obliquilobum *		Roadsides
TARAXACUM OBLONGATUM		Fertile damp meadows
TARAXACUM OSTENFELDII		Waste places
TARAXACUM OXONIENSE		Neutral or calcareous soils
Taraxacum pannucium *		Roadsides & waste places
TARAXACUM POLYODON		Roadsides & waste places
TARAXACUM PSEUDOHAMATUM		Roadsides, waste places & gardens
TARAXACUM QUADRANS		
TARAXACUM RUBICUNDUM		Dry calcareous download
Taraxacum stenacrum *.		Roadsides & waste places
Taraxacum subbracteatum *		
Taraxacum subexpallidum *		Roadsides & waste places
TARAXACUM SUBUNDULATUM		Water meadows & lush roadsides
TARAXACUM TAMASENSE		Wet hay meadows
Taraxacum tanyphyllum *		Road verges
Taraxacum undulatiflorum *		Roadsides & waste places
Taxodium distichum	Swamp Cypress	Very rare; by lakes in parkland; e.g. Black Park
TAXUS BACCATA	Yew	Locally common; woods; also frequently planted, especially in churchyards
TEESDALIA NUDICAULIS	Shepherd's-cress	NT; 1998 (Woburn Sands); dry sandy places
Telekia speciosa	Yellow Oxeye	Very rare; parkland & churchyard; e.g. Green Park
Tellima grandiflora	Fringecups	Very rare; woodland & hedges; e.g. Benhams Wood
TEPHROSERIS INTEGRIFOLIA SSP. INTEGRIFOLIA	Field Fleawort	EN; very rare; chalk grassland; e.g. Ivinghoe Beacon
Tetragonolobus maritimus	Dragon's-teeth	Very rare; roadsides; e.g. Fingest
Tetragonolobus purpureus	Asparagus-pea	2002 (Soulbury); old allotments
Teucrium chamaedrys *	Wall Germander	1960 (Aston Hill); calcareous grassland
TEUCRIUM SCORODONIA	Wood Sage	Scarce; dry open woods & grassy places on heathy soils
THALICTRUM FLAVUM	Common Meadow-rue	Rare; wet meadows, ditches & by rivers; decreasing
Thalictrum minus	Lesser Meadow-rue	Very rare; waste places; e.g. Wendover
THELYPTERIS PALUSTRIS	Marsh Fern	Very rare; boggy woodland at Great Brickhill
Thesium humifusum *	Bastard-toadflax	1805 (near Marlow); calcareous grassland
Thladiantha dubia	Manchu Tuber-gourd	1985 (unlocalised); waste ground
THLASPI ARVENSE	Field Penny-cress	Uncommon; arable fields & waste ground
Thlaspi perfoliatum *	Perfoliate Penny-cress	VU; 1957 (near Wendover); railway ballast
Thuja occidentalis	Northern White-cedar	1987 (Lane Wood); woodland
Thuja plicata	Western Red-cedar	Scarce; plantations
THYMUS POLYTRICHUS SSP. BRITANNICUS	Wild Thyme	Uncommon; short turf on porous soils; decreasing
THYMUS PULEGIOIDES	Large Thyme	Scarce; chalk grassland & heaths
Thymus vulgaris	Garden Thyme	1991 (Aylesbury); old allotments

TILIA CORDATA	Small-leaved Lime	Very rare; woods in the north-west; planted elsewhere
Tilia 'Petiolaris' *	Pendent Silver-lime	Pre-1926 (Stowe & Wharfe); parkland
Tilia platyphyllos ssp. cordifolia	Large-leaved Lime	Very rare; woods in the south; *e.g.* Fawley
Tilia tomentosa *	Silver-lime	Pre-1926 (Dropmore); parkland
Tilia x europaea	Common Lime	Common; parkland, hedges, woods & churchyards
Tolmiea menziesii	Pick-a-back-plant	Pre-1985 (Gerrards Cross)
Tordylium maximum *	Hartwort	1938 (near Bletchley); hedges
TORILIS ARVENSIS	Spreading Hedge-parsley	EN; very rare; arable fields on calcareous soils; *e.g.* Marlow Bottom
TORILIS JAPONICA	Upright Hedge-parsley	Very common; wood margins, hedgerows & field margins
TORILIS NODOSA	Knotted Hedge-parsley	Very rare; old short turf; especially in the north; decreasing
Trachyspermum ammi	Ajowan	1972 (Gerrards Cross); rubbish tip
Trachystemon orientalis	Abraham-Isaac-Jacob	Very rare; parkland & near gardens; *e.g.* Heelands
Tragopogon porrifolius	Salsify	Very rare; waste places; *e.g.* Haddenham
TRAGOPOGON PRATENSIS SSP. MINOR	Goat's-beard	Common; rough grassland, roadsides & meadows
Tragopogon pratensis ssp. orientalis *	Eastern Goat's-beard	Pre-1926 (near Hanslope); railway bank
Tragus racemosus *	European Bur-grass	1958 (Wooburn); paper mill waste
Trichophorum cespitosum ssp. germanicum *	Deer-grass	Pre-1926 (Stoke Common & Burnham Beeches); heaths
TRIFOLIUM ARVENSE	Hare's-foot Clover	Very rare; short turf & bare ground on light soils
Trifolium aureum	Large Trefoil	2002 (Tattenhoe); cultivated ground & grassland
TRIFOLIUM CAMPESTRE	Hop Trefoil	Uncommon; short old turf on dry soils
TRIFOLIUM DUBIUM	Lesser Trefoil	Common; grassland
TRIFOLIUM FRAGIFERUM SSP. FRAGIFERUM	Strawberry Clover	Scarce; roadsides & wet meadows
Trifolium hybridum ssp. hybridum	Alsike Clover	Uncommon; grassy places; possibly decreasing
Trifolium incarnatum ssp. incarnatum *	Crimson Clover	Pre-1926; fields & roadsides in the south
TRIFOLIUM MEDIUM	Zigzag Clover	Scarce; wood borders, hedges & railway banks in the south & north
TRIFOLIUM MICRANTHUM	Slender Trefoil	Rare; short turf & heaths
Trifolium pannonicum	Hungarian Clover	Very rare; railway bank at Forty Green, Beaconsfield
TRIFOLIUM PRATENSE	Red Clover	Very common; grassland
TRIFOLIUM REPENS	White Clover	Very common; grassland
Trifolium resupinatum *	Reversed Clover	Pre-1926 (near Eton); waste places
TRIFOLIUM STRIATUM	Knotted Clover	Very rare; dry sandy places; especially in the south
Trifolium subterraneum *	Subterranean Clover	1946 (Langley Park); dry gravelly places
TRIGLOCHIN PALUSTRE	Marsh Arrow-grass	Very rare; fens & wet grassland; especially in the north
Trigonella caerulea *	Blue Fenugreek	Pre-1926 (Iver); rubbish tip
Trigonella corniculata	Sickle-fruited Fenugreek	1970 (Iver); rubbish tip
Trigonella foenum-graecum *	Fenugreek	Pre-1926 (Iver & Slough); waste places
Trigonella procumbens *	Sheep's Clover	Pre-1926 (Langley, Eton & Slough); waste places
TRIPLEUROSPERMUM INODORUM	Scentless Mayweed	Common; rough places & arable fields
TRISETUM FLAVESCENS SSP. FLAVESCENS	Yellow Oat-grass	Common; grassland; especially on calcareous soils
Tristagma uniflorum	Spring Starflower	Very rare; churchyards & near habitation
Triticum aestivum	Bread Wheat	Common; roadsides & waste places
Triticum turgidum *	Rivet Wheat	Pre-1926 (unlocalised); waste places
Tropaeolum majus	Nasturtium	Rare; waste places
Tropaeolum peregrinum *	Canary-creeper	Pre-1926 (Iver); waste ground
Tsuga heterophylla	Western Hemlock-spruce	Rare; plantations; *e.g.* Bledlow Great Wood
Tulipa gesneriana	Garden Tulip	Rare; roadsides & waste places
Tulipa sylvestris	Wild Tulip	Very rare; roadside, parkland & churchyard; *e.g.* Addington
Turgenia latifolia	Greater Bur-parsley	1997 (Great Missenden); gardens & waste places
TUSSILAGO FARFARA	Colt's-foot	Common; roadsides, disturbed ground & by water
TYPHA ANGUSTIFOLIA	Lesser Bulrush	Scarce; ponds & rivers
TYPHA LATIFOLIA	Bulrush	Common; ponds, lakes, rivers & canals
TYPHA X GLAUCA		Very rare; water-filled pit at Bletchley
XTriticosecale	Triticale	1996 (Wraysbury); waste places
ULEX EUROPAEUS	Gorse	Common; heaths, commons & acid grassland

ULEX MINOR	Dwarf Gorse	Rare; heaths & open woods in the south
ULMUS GLABRA SSP. GLABRA	Wych Elm	Common; woods & hedges; especially in The Chilterns
Ulmus minor ssp. angustifolia*	Cornish Elm	Pre-1926; parkland; planted
ULMUS MINOR SSP. MINOR	Small-leaved Elm	Rare; woods & hedges
Ulmus minor ssp. sarniensis *	Jersey Elm	Pre-1926; parkland; planted
Ulmus plotii	Plot's Elm	1979 (Newton Longville); hedges
ULMUS PROCERA	English Elm	Very common; woods & hedges
Ulmus x hollandica	Dutch Elm	Uncommon; hedges
ULMUS X VEGETA	Huntingdon Elm	Rare; hedges
Umbilicus rupestris	Navelwort	Very rare; walls; e.g. Drayton Parslow
URTICA DIOICA SSP. DIOICA	Stinging Nettle	Very common; ubiquitous!
URTICA URENS	Small Nettle	Uncommon; waste places; often on lighter soils
UTRICULARIA AUSTRALIS	Bladderwort	Very rare; ponds on acid soils; e.g. Burnham Beeches
UTRICULARIA VULGARIS	Greater Bladderwort	Very rare; lake & old gravel pits; e.g. Hedgerley Green
Vaccaria hispanica *	Cowherb	Pre-1926; waste places & railways
VACCINIUM MYRTILLUS	Bilberry	Very rare; woodland edges on acid soils in the Brickhills
VALERIANA DIOICA	Marsh Valerian	Rare; fens
VALERIANA OFFICINALIS	Common Valerian	Frequent; damp woodland, wet grassland & dry calcareous grassland
Valeriana pyrenaica *	Pyrenean Valerian	1966 (Pitstone Hill)
VALERIANELLA CARINATA	Keeled-fruited Cornsalad	Uncommon; dry, open places & walls
VALERIANELLA DENTATA	Narrow-fruited Cornsalad	EN; very rare; cornfields on chalk
VALERIANELLA LOCUSTA	Common Cornsalad	Uncommon; dry, open places & walls
VALERIANELLA RIMOSA	Broad-fruited Cornsalad	EN; 2002 (Stokenchurch); cornfields
Verbascum blattaria	Moth Mullein	2004 (Wraysbury); waste ground & gravel pits
Verbascum densiflorum	Dense-flowered Mullein	Very rare; road & pathsides; e.g. Middle Weald
Verbascum lychnitis	White Mullein	1974 (Parmoor); waste ground
VERBASCUM NIGRUM	Dark Mullein	Locally common; roadsides & grassland, usually on chalk
Verbascum phlomoides	Orange Mullein	1998 (Denham); waste ground
Verbascum phoeniceum	Purple Mullein	1988 (Chesham); waste ground
VERBASCUM THAPSUS	Great Mullein	Frequent; dry grassland, walls, waste places & cleared woodland
Verbascum virgatum	Twiggy Mullein	1985 (Britwell); waste places & roadside
VERBASCUM X SEMIALBUM		1996 (Hambleden); roadside
Verbena bonariensis	Argentine Vervain	2003 (Lane End); waste places
VERBENA OFFICINALIS	Vervain	Uncommon; dry, open places on chalk
VERONICA AGRESTIS	Green Field-speedwell	Uncommon; cultivated ground, churchyards & gardens
VERONICA ANAGALLIS-AQUATICA	Blue Water-speedwell	Uncommon; pond & river banks
VERONICA ARVENSIS	Wall Speedwell	Common; walls & open ground
VERONICA BECCABUNGA	Brooklime	Common; mud at edge of rivers, streams & ditches
VERONICA CATENATA	Pink Water-speedwell	Uncommon; pond & river banks
VERONICA CHAMAEDRYS	Germander Speedwell	Very common; grassland & open woods
Veronica filiformis	Slender Speedwell	Common; lawns, churchyards, riverside meadows
VERONICA HEDERIFOLIA	Ivy-leaved Speedwell	Common; waste ground & shady places; both sspp. hederifolia & lucorum occur but distributions not known
VERONICA MONTANA	Wood Speedwell	Locally frequent; woods
VERONICA OFFICINALIS	Heath Speedwell	Uncommon; dry grassland on light soils
Veronica persica	Common Field-speedwell	Very common; cultivated ground & waste places
Veronica polita	Grey Field-speedwell	Uncommon; cultivated ground, churchyards & gardens
VERONICA SCUTELLATA	Marsh Speedwell	Very rare; ponds on heathy commons
VERONICA SERPYLLIFOLIA SSP. SERPYLLIFOLIA	Thyme-leaved Speedwell	Common; damp grassland & woodland rides
Veronica spicata	Spiked Speedwell	1986 (Cliveden); parkland
Veronica triphyllos *	Fingered Speedwell	RDB; 1871 (Langley); clover field
Veronica x lackschewitzii *		1968 (Missenden Abbey); pond
VIBURNUM LANTANA	Wayfaring-tree	Common; hedgerows & wood margins
VIBURNUM OPULUS	Guelder-rose	Common; hedgerows & woods
Viburnum rhytidophyllum	Wrinkled Viburnum	Very rare; planted in gardens; very rarely self-sown nearby
Viburnum tinus	Laurustinus	Rare; planted in parks & churchyards
Vicia benghalensis *	Purple Vetch	1928 (unlocalised); waste places
Vicia bithynica *	Bithynian Vetch	VU; 1956 (Burnham Beeches); waste places

VICIA CRACCA	Tufted Vetch	Common; hedgerows & wood margins
Vicia faba	Broad Bean	Uncommon; crop relic
VICIA HIRSUTA	Hairy Tare	Common; dry grassy places
Vicia hybrida *	Hairy Yellow-vetch	1897 (Chalfont); cultivated fields
VICIA LATHYROIDES	Spring Vetch	Very rare; dry sandy grassland; *e.g.* Rammamere Heath
Vicia lutea	Yellow-vetch	NT; 1998 (Chalfont St Peter); roadside
Vicia narbonensis *	Narbonne Vetch	Pre-1926 (Langley); waste ground
Vicia pannonica *	Hungarian Vetch	1941 (High Wycombe)
Vicia pannonica ssp. striata *		1903 (Iver Heath); clover field
Vicia parviflora	Slender Tare	VU; NS; 1992 (Heelands); dry grassy banks
VICIA SATIVA SSP. NIGRA	Narrow-leaved Vetch	Uncommon; grassy banks
Vicia sativa ssp. sativa		Very rare; relic of cultivation
VICIA SATIVA SSP. SEGETALIS	Common Vetch	Very common; grassy places
VICIA SEPIUM	Bush Vetch	Common; woodland & hedgebanks
VICIA SYLVATICA	Wood Vetch	Very rare; woodland on chalk
Vicia tenuifolia *	Fine-leaved Vetch	1954 (Wycombe); waste places
VICIA TETRASPERMA	Smooth Tare	Common; dry grassy places
Vicia villosa	Fodder Vetch	1992 (Willen); waste places & grassland
Vinca major	Greater Periwinkle	Uncommon; hedgebanks near habitation
Vinca minor	Lesser Periwinkle	Uncommon; hedgebanks & woods
VIOLA ARVENSIS	Field Pansy	Common; cultivated ground & waste places
VIOLA CANINA SSP. CANINA	Heath Dog-violet	NT; very rare; damp meadows
Viola canina x lactea *		1927 (Dropmore); heathy ground
VIOLA HIRTA	Hairy Violet	Common; dry calcareous grassland
Viola lactea *	Pale Dog-violet	1950 (Egypt); heathy ground
VIOLA ODORATA	Sweet Violet	Common; hedgerows, churchyards & gardens
VIOLA PALUSTRIS SSP. PALUSTRIS	Marsh Violet	Very rare; boggy ground on acid soil
VIOLA REICHENBACHIANA	Early Dog-violet	Locally common; woods
VIOLA RIVINIANA	Common Dog-violet	Common; woods & hedges
VIOLA TRICOLOR SSP. TRICOLOR	Wild Pansy	NT; rare; cornfields & waste ground; often a garden escape
Viola x bavarica *		1904 (Angling Spring & Amersham); woods
Viola x intersita *		1904 (Mop End); roadside
Viola x mixta *		Pre-1926 (Ashridge)
Viola x scabra *		1974 (Denbigh Hall); roadside
Viola x wittrockiana	Garden Pansy	Uncommon; waste places & rubbish tips
VISCUM ALBUM	Mistletoe	Locally common; mostly on trees in the south

MISTLETOE

Vitis vinifera	Grape-vine	Very rare; waste places
VULPIA BROMOIDES	Squirreltail Fescue	Rare; waste ground
Vulpia ciliata ssp. ambigua	Bearded Fescue	1984 (Weston Underwood); wall
VULPIA MYUROS	Rat's-tail Fescue	Scarce; waste ground
Vulpia unilateralis *	Mat-grass Fescue	1964 (Dancersend); chalk grassland
Weigelia florida	Weigelia	1982 (Stone)
Xanthium spinosum *	Spiny Cocklebur	Pre-1926; waste places in the south
Xanthium strumarium *	Rough Cocklebur	1912 (Uxbridge); waste ground
Yucca gloriosa	Spanish-dagger	1999 (Beaconsfield); waste ground
ZANNICHELLIA PALUSTRIS	Horned Pondweed	Rare; scattered in ponds & shallow streams
Zea mays	Maize	Very rare; crop relic

Species recorded in error

Alchemilla monticola (Stowe Park); *Andromeda polifolia* (Iver Heath, 1805); *Astragalus danicus* (Chiltern Open Air Museum, 1993); *Callitriche hermaphroditica* (Linford Pits, 1973); *Callitriche palustris* (Halton, pre-1926; Rushbeds Wood, 1969); *Campanula patula* (Medmenham Hill, 1857 - for *C. rapunculus*); *Cardamine impatiens* (unlocalised, pre-1926 - for *C. flexuosa*); *Carex dioica* (Fivearch Wood, pre-1978; Rushbeds Wood, pre-1985); *Carex divisa* (Wootton, pre-1978, Windsor Hill, 1987); *Carex elongata* (Slough, 1940 - for *C. paniculata*); *Carex x boenninghausiana* (unlocalised); *Centaurium littorale* (several records - for *C. erythraea*); *Cerastium pumilum* (West Wycombe Hill, 1986); *Chrysosplenium alternifolium* (Cliveden, 1835; Wootton Lakes, 1985); *Consolida orientalis* (unlocalised - for *C. regalis*); *Corynephorus canescens* (Agars Plough); *Crepis foetida* (nr Bulstrode Park, 1780; Brill Common, 1995); *Crocus flavus* (Loughton - for *C. x stellaris*); *Deschampsia setacea* (Burnham Beeches, 1950); *Diplotaxis erucoides* (Hoppers Field, 1986); *Drosera anglica* (Burnham Beeches, 1954); *Egeria densa* (Dinton Hall, 1981); *Eleocharis uniglumis* (Hollington, 1996; Great Brickhill, 1996); *Elodea callitrichoides* (Jubilee Pit, pre-1985 - for *Lagarosiphon major*); *Euphorbia esula* (unlocalised - for *E. x pseudovirgata*); *Fragaria chiloensis* (Castlethorpe, pre-1926; Wendover, pre-1926 - for *F. x ananassa*); *Gentianella campestris* (Ellesborough; Monks Risborough; Keep Hill; all pre-1926); *Hieracium diaphanum* (Hanslope, 1961); *Hieracium lepidulum* (Gerrards Cross, 1976); *Illecebrum verticillatum* (unlocalised); *Impatiens noli-tangere* (Burnham Beeches, 1950 - for *I. parviflora*); *Ledum palustre* ssp. *groenlandicum* (Oak Wood, 1981); *Legousia speculum-veneris* (unlocalised, pre-1970 - for *L. hybrida*); *Leucanthemum maximum* (several records - for *L. x superbum*); *Listera cordata* (unlocalised, 1872); *Lobelia urens* (unlocalised, pre-1986) *Lunaria rediviva* (several records - for *L. annua*); *Lupinus polyphyllus* (several records - for *L. x regalis*); *Melampyrum sylvaticum* (unlocalised); *Mentha longifolia* (several records - for *M. spicata*); *Mimulus luteus* (Marlow, 1959 - for *M. guttatus*); *Orchis purpurea* (Downley; Deangarden Wood; Fennell's Wood; all pre-1926 - for *O. militaris*); *Parapholis incurva* (Iver, 1903 - for *P. pycnantha*); *Pilosella x stoloniflora* (Hanslope, 1899 - for *P. caespitosa* ssp. *colliniformis*); *Potamogeton coloratus* (Scotsgrove, pre-1985); *Potamogeton gramineus* (Great Linford, 1981 - for *P. natans*); *Pyrola media* (Wendover, 1835 - for *P. minor*); *Pyrola rotundifolia* (unlocalised - for *P. minor*); *Ranunculus omiophyllus* (Farnham Common, pre-1877); *Ranunculus sphaerospermus* (Halton, 1914 - for *R. penicillatus* ssp. *pseudofluitans*); *Ranunculus tripartitus* (Gerrards Cross Common, 1954); *Rhus coriaria* (Sands, 1985; Notley Abbey, pre-1985); *Rosa caesia* ssp. *caesia* (Naphill Common, 1917); *Rosa sherardii* (several records; all pre-1926); *Rumex longifolius* (unlocalised, pre-1986); *Salix babylonica* (several records); *Salix myrsinifolia* (Alderbourne Bottom, pre-1926); *Salix phylicifolia* (Alderbourne Bottom, pre-1926; Pens Place, 1986); *Sedum anglicum* (unlocalised); *Sedum stoloniferum* (Buckingham, pre-1926 - for *S. spurium*); *Senecio paludosus* (Grendon & Doddershall Woods, 1981 - for *S. aquaticus*); *Silene conica* (Homefield Wood, 1976); *Sorbus x vagans* (Wapsey's Wood, 1976 - for *S. intermedia*); *Spiraea salicifolia* (several records); *Stachys germanica* (unlocalised, pre-1953); *Stellaria nemorum* (South Bucks, 1869); *Taraxacum duplidentifrons* (unlocalised, pre-1985); *Taraxacum maculatum* (unlocalised, pre-1972); *Taraxacum tenebricans* (unlocalised, pre-1986); *Trifolium ochroleucon* (Woburn Sands, 1981); *Trifolium ornithopodioides* (unlocalised); *Ulex gallii* (several records; all pre-1877 - for *U. minor*); *Utricularia intermedia* (Stoke Common, 1861, Burnham Beeches, pre-1877 - for *U. australis*); *Vicia orobus* (Wendover Forest, 1967); *Viola benghalense* (unlocalised - for *Vicia benghalensis*); *Viola rupestris* (unlocalised, pre-1932 - for *V. riviniana*); *Wolffia arrhiza* (Weston Turville, 1972).

Species requiring confirmation

Aesculus parviflora (unlocalised, pre-1986); *Agrostis curtisii* (Stoke Common, 1970); *Apium repens* (Datchet Common, 1805); *Barbarea stricta* (Kingsey, pre-1985); *Carex viridula* ssp. *viridula* (Shabbington Wood, 1981); *Carum verticillatum* (nr Seer Green, 1894); *Centaurea jacea* (nr Kit's Wood, 1987); *Centaurea x moncktonii* (SU99, 1970); *Centaurium pulchellum* (Church Wood, 1976; Butterleys Wood, 1982); *Commelina coelestis* (Iver, 1972); *Dryopteris x uliginosa* (Little Brickhill, pre-1926; Widdenton Park Wood, 1867); *Euphorbia coralloides* (nr Taplow Common, 1904); *Euphrasia stricta* (nr Seer Green, 1898); *Fumaria officinalis x densiflora* (nr Wendover, pre-1926); *Hammarbya paludosa* (East Burnham Common, pre-1877); *Hieracium acuminatum* (Bletchley, 1997); *Hieracium pellucidum* (Bulstrode, 1898); *Hieracium sublepistoides* (Aston Clinton, 1956); *Hordeum marinum* (unlocalised, pre-1953); *Isolepis cernua* (unlocalised, 1904); *Lavatera arborea* (unlocalised, pre-1985); *Linum perenne* ssp. *anglicum* (Ludgershall, 1978); *Lithospermum purpureocaeruleum* (Chetwode Manor, 1997); *Lotus angustissimus* (Hedgerley, 2002); *Lycium chinense* (unlocalised, pre-1926); *Ononis x pseudohircina* (Denbigh Hall, 1974; Newlands, 1974; Two Mile Ash, 1974); *Orchis simia* (nr Ibstone, pre-1926); *Pilularia globulifera* (unlocalised, pre-1926); *Plantago lanceolata x media* (Longwick, pre-1926); *Polygonatum odoratum* (Dropmore, 1961); *Potamogeton x grovesii* (Marsh Gibbon, pre-1926); *Potentilla palustris* (nr Marlow, 1868); *Ranunculus baudotii* (Stoke Poges, 1862); *Ranunculus penicillatus* ssp. *penicillatus* (River Colne, 1984); *Rosa caesia* ssp. *vosagiaca* (nr Halton; nr Horton; Swanbourne; all pre-1926); *Rosa x verticillacantha* (several records, pre-1926; Salcey Forest, 1992); *Rumex pseudoalpinus* (Missenden Abbey, 1996); *Sisymbrium loeselii* (unlocalised, pre-1986) *Sisyrinchium bermudianum* (unlocalised, pre-1986) *Symphytum asperum* (nr Chalvey, 1982); *Taraxacum bracteatum* (nr. Kimble, 1921); *Taraxacum dahlstedtii* (Hanslope, pre-1926); *Taraxacum hamatiforme* (unlocalised, pre-1972); *Taraxacum lamprophyllum* (unlocalised, pre-1986); *Taraxacum lingulatum* (unlocalised, pre-1986); *Taraxacum longisquameum* (nr Princes Risborough, 1914); *Taraxacum sellandii* (unlocalised, pre-1986); *Taraxacum sublaeticolor* (unlocalised, pre-1986); *Taraxacum unguilobum* (Bletchley, 1914); *Tragopogon pratensis* ssp. *pratensis* (College Lake, 1992); *Trifolium scabrum* (Iver, 1964); *Utricularia minor* (Uxbridge River, 1746; Burnham Beeches, 1931); *xFestulolium brinkmannii* (Well End, pre-1985).

Species listed in "Vice County Census Catalogue" for which no record has been traced

Anthyllis vulneraria ssp. *polyphylla*; *Avena fatua x sativa*; *Epilobium x rivulare*; *Epilobium x semiobscurum*; *Euphrasia nemorosa x pseudokerneri*; *Euphrasia x haussknechtii*; *Ferula communis*; *Fuchsia magellanica*; *Leontodon hispidus x saxatilis*; *Lonicera x italica*; *Persicaria campanulata*; *Petrorhagia prolifera*; *Rumex x pseudopulcher*; *Spiraea x rosalba*; *Ulmus x elegantissima*.

REFERENCES

Burton, R.M. (1983). Flora of the London Area. London Natural History Society

Cheffings, C.M. & Farrell, L. (Eds), Dines, T.D., Jones, R.A., Leach, S.J., McKean, D.R., Pearman, D.A., Preston, C.D., Rumsey, F.J. & Taylor, I. (2005). The Vascular Red Data List for Great Britain. *Species Status,* **7**:1-116. JNCC, Peterborough.

Druce, G.C. (1926). The Flora of Buckinghamshire. Buncle. Arbroath.

Dudman, A.A. & Richards, A.J. (1997). Dandelions of Great Britain and Ireland. BSBI, London.

Graham, G.G. & Primavesi, A.L. (1993). Roses of Great Britain & Ireland. BSBI., London.

Maycock, R. (1985). The Flora of Buckinghamshire Churchyards. M.Sc. thesis, University of Durham.

Maycock, R. (1986). List of the Vascular Plants of Buckinghamshire. Unpublished.

Maycock, R. & Woods, A. (2000). in Milton Keynes More Than Concrete Cows, real animals and plants too. Milton Keynes Natural History Society.

Moyes, N.J. & Willmot, A. (2002). A Checklist of the Plants of Derbyshire. Derby Museum.

Newton, A. & Edees, E.S. (1988). Brambles of the British Isles. The Ray Society, London.

Newton, A. & Randall, R.D. (2004). Atlas of British and Irish Brambles. BSBI., London.

Stace, C.A. (1997). New Flora of the British Isles (2nd edition). Cambridge.

Stace, C.A., Ellis, R.G., Kent, D.H. & McCosh, D.J. (2003). Vice-County Census Catalogue of the Vascular Plants of Great Britain. BSBI. London

Preston, C.D., Pearman, D.A. & Dines, T.D. (2002). New Atlas of the British & Irish Flora. Oxford.

Richards, A. J. (1972), The *Taraxacum* Flora of the British Isles. *Watsonia*, **9** (supplement)

Stewart, A., Pearman, D.A. & Preston, C.D. (1994). Scarce Plants in Britain. JNCC. Peterborough.

Wiggington, M.J. (1999). British Red Data Book Vascular Plants (3rd edition). JNCC. Peterborough.

VICE-COUNTY 24 (BUCKS)

Turweston
Biddlesden
Water Stratford
Stowe
Lillingstone Lovell
Maids Moreton
Leckhamstead
Tingewick
Buckingham
Thornborough
Twyford
Calvert
Marsh Gibbon
Middle Claydon
Grendon Underwood
Ludgershall
Wotton Underwood

Great Horwood
Adstock
Little Horwood
Addington
Winslow
Swanbourne
Whaddon
Mursley
Drayton Parslow
Stewkley

Hanslope
Castlethorpe
Stony Stratford
Calverton
Wolverton
Great Linford
Stantonbury
Downs Barn
Heelands
Bradwell
Springfield
Milton Keynes
Woolstone
Campbell Park
Loughton
Willen
Broughton

Lathbury
Newport Pagnell
Tongwell

Lavendon
Cold Brayfield
Newton Blossomville
Olney
Weston Underwood
Emberton
Stoke Goldington

North Crawley
Mousloe

Wavendon
Wavendon Gate
Woburn Sands
Caldecotte
Bow Brickhill
Fenny Stratford
Bletchley
Water Eaton
Little Brickhill
Great Brickhill
Stoke Hammond
Soulbury

Burcott
Wing
Cublington
Whitchurch
Aston Abbotts
Hardwick
Wingrave
Weedon

Northall
Edlesborough
Mentmore
Cheddington
Ivinghoe

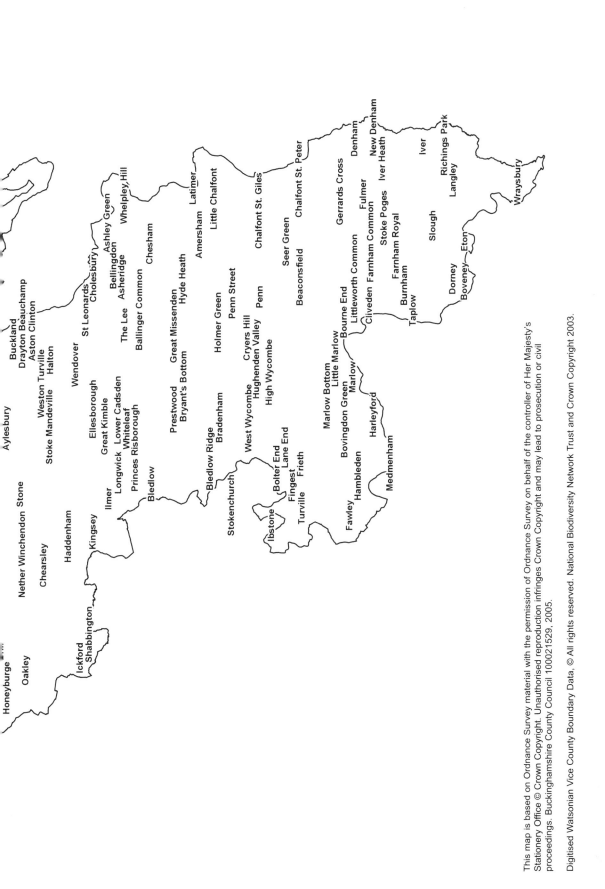

This map is based on Ordnance Survey material with the permission of Ordnance Survey on behalf of the controller of Her Majesty's Stationery Office © Crown Copyright. Unauthorised reproduction infringes Crown Copyright and may lead to prosecution or civil proceedings. Buckinghamshire County Council 100021529, 2005.

Digitised Watsonian Vice County Boundary Data, © All rights reserved. National Biodiversity Network Trust and Crown Copyright 2003.

GAZETTEER

Each place name is followed by a 4-figure grid reference so that it can be located within a 1km square. The naming of a place on the map or within the text does not indicate public access so, if in any doubt, ensure that permission is obtained before entering any site.

Place	Ref	Place	Ref	Place	Ref
Amerden	SU9080	Hell Coppice	SP6010	Sands Bank	SU8493
Angling Spring Wood	SP8801	Hillock Wood	SP8302	Sawyers Green	TQ0180
Ashridge	SP9713	Hockeridge Wood	SP9706	Shenley Church End	SP8237
Aston Ferry	SP7884	Hollywell Plantation	SP6236	Simpson	SP8836
Aston Hill	SP8810	Horton Wharf	SP9319	Singleborough	SP7632
Bacombe Hill	SP8606	Howe Park Wood	SP8334	Sneshall East	SP8232
Bakers Wood	TQ0287	Hughenden	SU8694	Spade Oak	SU8887
Ballards Wood	SP8909	Hunstmoor Park	TQ0482	Splash Covert	SP8611
Ballinger Dell Wood	SP9103	Hyde Lane	SP7235	Stanton Low	SP8342
Barn Wood	SP8806	Joan's Piece	SP8442	Startop's End	SP9114
Benham Wood	SU7685	Jubilee Pit	SP8630	Stockgrove	SP9129
Black Park	TQ0184	Kilwick Wood	SP8653	Stoke Common	SU9885
Blackend Spinney	SP8625	King's Wood	SU8993	Stonebridge	SP8214
Bockmer End	SU8186	Lamport Wood	SP6837	Stonepit Field	SP8542
Booker	SU8391	Lane Wood	SU9898	Sutton	TQ0278
Boswells	SP9706	Langley Park	TQ0179	Tathall End	SP8248
Bow Wood	SU9895	Lenborough Wood	SP6631	Tattenhoe	SP8333
Bragenham	SP9028	Linford Pits	SP8442	Temple Island	SU7784
Brands Hill Park	SU8794	Linford Wood	SP8440	Thorney	TQ0479
Britwell	SU9482	Linslade	SP9025	Tickford End	SP8843
Broomhills Wood	SP9132	Little Hampden	SP8503	Tiddenfoot Water Park	SP9123
Broughton	SP8940	Little Missenden	SP9298	Totteridge	SU8893
Bulstrode Park	SU9888	Lodge Wood	SP8601	Upper Hartwell	SP7812
Burnham Beeches	SU9585	Loudwater	SU9090	Walk Wood	TQ0098
Chalfont Grove	SU9891	Low Scrubs	SP8506	Walters Ash	SU8398
Chenies	TQ7110	Lower End, Wavendon	SP9137	Well End	SP8118
Chesham Bois	SU9699	Lucas Wood	SU8793	Wexham	SU9983
Chesham Moor	SP9600	Micklefield	SU8892	Whitfield Wood	SP6439
Church Wood	SU9787	Middle Green	TQ0080	Windsor Hill	SP8202
Cippenham	SU9480	Middle Weald	SP7938	Widmer End	SU8896
Claydon Woods	SP7123	Milton Keynes Village	SP8839	Wilstone Reservoir	SP9013
Clifton Reynes	SP9051	Moor Wood	SU8190	Windmill Plantation	SU9598
Cliveden	SU9185	Moorend Common	SU8090	Wooburn	SU9087
Cold Brayfield	SP9252	Mop End	SU9297	Woughton Park	SP8836
College Lake	SP9314	Moreton Green	SU8407	Wycombe Marsh	SU8891
Colnbrook	TQ0277	Naphill	SU8497		
Common Wood	SU9194	New Bradwell	SP8241		
Commonhill Wood	SU7594	New Wavendon Heath	SP9234		
Combe Hill	SU9087	Newland Park	TQ0193		
Conniburrow	SP8539	Old Linslade	SP9126		
Cowcroft	SP9801	Old Rectory Wood	TQ0387		
Cub Pond	SP6519	Old Slade	TQ0387		
Dancersend	SP9009	Old Wavendon Heath	SP9334		
Danesfield	SU8184	Old Wolverton	SP8041		
Davenport Wood	SU8286	Olney Park	SP8753		
Denbigh Hall	SP8535	Pancake Wood	SP9706		
Dibden Hill	SU9992	Park Wood	SU8298		
Ditton Park	SU9978	Parmoor	SU7889		
Dorney Wood	SU9485	Partridge Hill	SP8929		
Downley	SU8494	Pens Place	SU8485		
Dropmore	SU9288	Peterley Wood	SU8799		
East Burnham Common	SU9584	Pheasants Hill	SU7887		
Egypt	SU9586	Philipshill Wood	TQ0194		
Eton Wick	SU9478	Pilch Fields	SP7432		
Fawley Bottom	SU7486	Pink Hill	SU8201		
Fennells Wood	SU8990	Pitstone	SP9315		
Finemere Wood	SP7121	Prestonn Bissett	SP6529		
Four Ashes	SU8795	Pulpit Hill	SP8304		
Grangelands	SP8304	Rammamere Heath	SP9230		
Great Kingshill	SU8797	Ravenstone	SP8450		
Great Tinkers Wood	SU8594	Rowley Wood	SU9983		
Green Park	SP8811	Rowsham	SP8417		
Grubbins Plantation	SU8998	Rush Green	TQ0385		
Handy Cross	SU8590	Rushmere Park	SP9128		
Hazeley Wood	SP8137	Salcey Forest	SP8150		
Hedgerley	SU9687	Salden	SP8229		
Hedgerley Green	SU9787	Salt Hill	SU9680		